Value-Distribution Theory

(in two parts)

Part B

PURE AND APPLIED MATHEMATICS

A Series of Monographs and Textbooks

1. K. Yano. Integral Formulas in Riemannian Geometry (1970)
2. S. Kobayashi. Hyperbolic Manifolds and Holomorphic Mappings (1970)
3. V. S. Vladimirov. Equations of Mathematical Physics (A. Jeffrey, editor; A. Littlewood, translator) (1970)
4. B. N. Pshenichnyi. Necessary Conditions for an Extremum (L. Neustadt, translation editor; K. Makowski, translator) (1971)
5. L. Narici, E. Beckenstein, and G. Bachman. Functional Analysis and Valuation Theory (1971)
6. D. S. Passman. Infinite Group Rings (1971)
7. L. Dornhoff. Group Representation Theory (in two parts). Part A: Ordinary Representation Theory. Part B: Modular Representation Theory (1971, 1972)
8. W. Boothby and G. L. Weiss (eds.). Symmetric Spaces: Short Courses Presented at Washington University (1972)
9. Y. Matsushima. Differentiable Manifolds (E. T. Kobayashi, translator) (1972)
10. L. E. Ward, Jr. Topology: An Outline for a First Course (1972)
11. A. Babakhanian. Cohomological Methods in Group Theory (1972)
12. R. Gilmer. Multiplicative Ideal Theory (1972)
13. J. Yeh. Stochastic Processes and the Wiener Integral (1973)
14. J. Barros-Neto. Introduction to the Theory of Distributions (1973)
15. R. Larsen. Functional Analysis: An Introduction (1973)
16. K. Yano and S. Ishihara. Tangent and Cotangent Bundles: Differential Geometry (1973)
17. C. Procesi. Rings with Polynomial Identities (1973)
18. R. Hermann. Geometry, Physics, and Systems (1973)
19. N. R. Wallach. Harmonic Analysis on Homogeneous Spaces (1973)
20. J. Dieudonné. Introduction to the Theory of Formal Groups (1973)
21. I. Vaisman. Cohomology and Differential Forms (1973)
22. B.-Y. Chen. Geometry of Submanifolds (1973)
23. M. Marcus. Finite Dimensional Multilinear Algebra (in two parts) (1973)
24. R. Larsen. Banach Algebras: An Introduction (1973)
25. R. O. Kujala and A. L. Vitter (eds.). Value Distribution Theory: Part A; Part B, Deficit and Bezout Estimates by Wilhelm Stoll (1973)

In Preparation:

K. B. Stolarsky. Algebraic Numbers and Diophantine Approximation

Value-Distribution Theory

(in two parts)

Proceedings of the Tulane University Program on Value-Distribution Theory in Complex Analysis and Related Topics in Differential Geometry

Edited by

Robert O. Kujala and Albert L. Vitter III

Department of Mathematics
Tulane University
New Orleans, Louisiana

Part B

Deficit and Bezout Estimates

Wilhelm Stoll

Department of Mathematics
Notre Dame University
Notre Dame, Indiana

MARCEL DEKKER, INC. New York 1973

MARCEL DEKKER, INC.
95 Madison Avenue, New York, New York 10016

LIBRARY OF CONGRESS CATALOG CARD NUMBER: 73-89281
ISBN: 0-8247-6125-1

Printed in the United States of America

CONTENTS

EDITORS' FOREWORD

We wish to take this opportunity to thank all those persons who contributed to the success of the Tulane University Semester Program in Value-Distribution Theory and more particularly to the production of this volume, especially:

Professor Wilhelm Stoll, the extent of whose participation in our program is only partially reflected by the contents of this volume since it does not indicate, for example, the amount of time which he devoted to conversations with the faculty, students, and other participants in the program during his visit here nor the fact of his personal supervision of all aspects of the actual preparation of this publication;

The members of the Department of Mathematics and the administration at the University of Notre Dame who generously permitted Professor Stoll's leave of absence for his visit;

Our colleagues in the Department of Mathematics here whose sympathy with the goals of this program made possible the use of a portion of the funds awarded to the University in a Science Development Grant Renewal from the National Science Foundation for the organization of this program;

The other invited participants in the program, the fruits of whose visits are represented by their contributions in Part A of these proceedings;

Gloria Montufar of the staff of the Department of Mathematics at the University of Notre Dame who did the actual typing of this manuscript thereby permitting the quick production of this volume and earning the eternal gratitude of the secretarial staff here at Tulane.

Robert O. Kujala

Albert L. Vitter III

Tulane University
New Orleans, Louisiana

PREFACE

During the Spring Semester 1973, I was a visiting professor at Tulane University to participate in a Special Semester on value distribution in several complex variables and to give a course on my research in this subject matter. The first few weeks were devoted to the average Bezout estimate [45]. Then some results of Griffiths [16] were presented. Soon it became apparent that they could be extended to study the intersections of a pure dimensional analytic subset M of a complex vector space V with the linear subspaces of codimension one.

The idea was to apply the method of [36] to a non-harmonic exhaustion and to use the Plücker difference formula and the Ahlfors estimates to obtain an estimate of the deficit term in the First Main Theorem. This can be applied to obtain Bezout estimates.

So, while this program of research was being carried out, I lectured currently on these investigations for the remaining three months.

Part of the results could be carried over to complex spaces. But to avoid purely technical difficulties, the theory is developed on manifolds only. For general references, the reader is referred once and for all to Ahlfors [1], Weyl [48], Wu [51] and to [36].

Some of the main results are explained in the introduction.

In section 2, some algebraic and differential geometric features associated to hermitian vector spaces are assembled. In section 3, the Stokes Theorem for complex spaces is stated in the formulation of Tung [47]. Although mostly complex manifolds are considered, it is handy to have the Stokes theorem available on complex spaces. Meromorphic maps are defined in section 4 and the connection to the hypersection bundle is established. Section 5 brings the First Main Theorem, which is proved using the Gauss-Bonnet formula. A first step to the Ahlfors estimates is made by integral averaging in section 6. The Hölder inequality is used to estimate the remainder terms. In section 7, contravariant differentiation is studied, which leads to the associated maps. Special care is taken to prove the existence of such a differentiation under reasonable assumptions. The Plücker difference formula is derived in section 8. Under the assumption that the covariant vector field is majorized by the (2m - 2)-dimensional measure in M, the Ahlfors estimates are established in section 9 by an averaging method. In section 10, pseudoconvex manifolds are considered. The abstract theory becomes concrete. An important discovery is the independence of the value distribution functions from a convex change of scale of the exhaustion function. Finally, the first deficit estimates are obtained in section 11. Uniform and asymptotic estimates are derived. In section 12, the characteristic functions of the associated maps are eliminated by the use of the Plückert Difference Formula. In section 13, a radical assumption is made which controls the inflection. The third deficit estimates are obtained. Section 14 applies all these results to the Bezout problem. The general assumptions are collected in a special index at the end of the paper.

All these investigations are directed towards a Bezout estimate. Obviously the next step will be to attempt a defect relation in this case of a non-harmonic exhaustion.

This monograph is a research paper as well as the set of notes lectured to my class at Tulane. The University of Notre Dame granted me leave of absence to participate at the Conference. The research was supported by Tulane University and the National Science Foundation as a sponsor of the conference and by its research grant NSF GP20139. I thank these organizations for their help and support which made this work possible.

Wilhelm Stoll
Tulane University
Spring 1973

GERMAN LETTERS

A	B	C	D	E	F	G	H	I	J	K	L	M	N	O	P

Q	R	S	T	U	V	W	X	Y	Z

a	b	c	d	e	f	g	h	i	j	k	l	m	n	o	p

q	r	s	t	u	v	w	x	y	z

CONTENTS OF PART A

DEFICIT AND BEZOUT ESTIMATES

I. INTRODUCTION

Let V be a complex vector space of dimension n+1 with $n > 0$. Let V^* be the dual vector space. The complex projective space $\mathbb{P}(V)$ of V is given. The natural projection

$$\mathbb{P}: V - \{0\} \to \mathbb{P}(V)$$

is defined such that $\mathbb{P}^{-1}(\mathbb{P}(\mathfrak{x})) = \{z\mathfrak{x} \mid 0 \neq z \in \mathbb{C}\}$. For $A \subseteq \mathbb{P}(V)$ define $\mathbb{P}(A) = \mathbb{P}(A - \{0\})$. Given $a \in \mathbb{P}(V^*)$, take any $\alpha \in \mathbb{P}^{-1}(a)$. Then $\alpha: V \to \mathbb{C}$ is linear. Therefore the linear subspace $E[a] = \ker \alpha$ has dimension n and does not depend on the choice of α. The projective space $\ddot{E}[a] = \mathbb{P}(E[a])$ is embedded in $\mathbb{P}(V)$ and is called a hyperplane of $\mathbb{P}(V)$.

Let M be a connected, non-compact, complex manifold of dimension $m > 0$. Consider a meromorphic map

$$f: M \longrightarrow \mathbb{P}(V)$$

such that $f(M) \not\subseteq \ddot{E}[a]$ for all $a \in \mathbb{P}(V^*)$. The growth of the inverse images $f^{-1}(\ddot{E}[a])$ shall be investigated. Obviously, adequate growth measures have to be introduced.

On V, take a positive definite hermitian form $(\mathfrak{z}|\mathfrak{w})$ as a hermitian product. Define $|\mathfrak{z}| = \sqrt{(\mathfrak{z}|\mathfrak{z})}$. Define $\tau_0 : V \to \mathbb{R}$ by $\tau_0(\mathfrak{z}) = |\mathfrak{z}|^2$. Let $d = \partial + \bar{\partial}$ be the exterior derivative which is twisted to

$$d^{\perp} = i(\partial - \bar{\partial}) = - d^c.$$

On V, differential forms

$$\upsilon = \frac{1}{4\pi} dd^c \tau_0 > 0$$

$$\omega = \frac{1}{4\pi} dd^c \log \tau_0 \geqq 0$$

are defined. The hermitian product on V defines the Fubini-Study Kaehler metric with fundamental form $\ddot{\omega}$ on $\mathbb{P}(V)$ such that

$$\mathbb{P}^*(\ddot{\omega}) = \omega$$

is the pullback. The hermitian product on V induces hermitian products on V^* and $\underset{p}{\Lambda} V$ for $p = 0,1,\ldots,n$. The corresponding data are denoted by $\tau_0, \upsilon, \omega, \ddot{\omega}$ for V^* and $\tau_0, \upsilon_p, \omega_p, \ddot{\omega}_p$ for $\underset{p+1}{\Lambda} V$.

Given $z \in \mathbb{P}(V)$ and $a \in \mathbb{P}(V^*)$, take any $\mathfrak{z} \in \mathbb{P}^{-1}(z)$ and $\alpha \in \mathbb{P}^{-1}(a)$, then the <u>distance</u>

$$0 \leqq \| z;a \| = \frac{|\alpha(\mathfrak{z})|}{|\alpha||\mathfrak{z}|} \leqq 1$$

is well defined.

If $a \in \mathbb{P}(V)$, then $f^{-1}(\ddot{E}[a]) = \gamma_f(a)$ is either empty or has pure dimension m-1. Here $\gamma_f(a)$ has to be counted "with multiplicities", i.e., $\gamma_f(a)$ has to be regarded as the support of some <u>intersection divisor</u> δ_f^a. Take $x_0 \in M$. Then an open neighborhood U of x_0 and a holomorphic vector function $\mathfrak{w}: U \to V$ exists such that

$$\dim \mathfrak{w}^{-1}(0) \leqq m-2$$

$$f = \mathbb{P} \circ \mathfrak{w} \quad \text{on } U - \mathfrak{w}^{-1}(0)$$

Such a vector function is called an irreducible representation of f on U. Then $U \cap \gamma_f(a) = (\alpha \circ \mathfrak{w})^{-1}(0)$. Now $\delta_f^a | U$ is given and well defined as the zero-divisor of the holomorphic function $\alpha \circ \mathfrak{w}$ on U. Obviously, δ_f^a is a non-negative divisor on M with $\gamma_f(a)$ as its support.

On M, a (2m-2)-dimensional measure is needed to measure the size of the divisor δ_f^a. So, it is assumed that a non-negative differential form χ of class C^∞ and bidegree (m-1, m-1) on M is given to provide this measure. Assume $d\chi = 0$ in addition. Here $\delta_f^a \chi$ may give an infinite measure to $\gamma_f(a)$. Hence a cut-off is needed.

A <u>bump</u> (G, g, ψ) is given by two relative compact open subsets G and g with C^∞-boundaries dG and dg and by a continuous function ψ such that $G \supset \overline{g}$ and such that $\psi | \overline{G} - g$ has class C^∞. Also $\psi | (M-G) = 0$ and $\psi | \overline{g} = R > 0$ are assumed to be constant with $0 \leqq \psi \leqq R$ on M. Here $g = \emptyset$ is permitted.

A divisor ν on M can be defined as a function $\nu: M \to \mathbb{Z}$ such that ν is locally given as the difference of the zero-divisors of two local holomorphic functions. The support $\gamma(\nu)$ is empty or a pure (m-1)-dimensional analytic subset of M. Therefore the <u>counting function</u>

$$n_\nu(G) = \int_{G \cap \gamma(\nu)} \nu\chi$$

and the <u>valence function</u>

$$N_\nu(G) = \int_{G \cap \gamma(\nu)} \nu\psi\chi$$

are defined. If $\nu \geqq 0$, then $n_\nu \geqq 0$ and $N_\nu \geqq 0$. If $a \in \mathbb{P}(V^*)$ and $\nu = \delta_f^a$ define

$$n_f(G,a) = n_\nu(G) \qquad\qquad N_f(G,a) = N_\nu(G)$$

The pullback of the differentials $d^\perp \psi \wedge \chi$ to the boundaries dG and dg is non-negative. <u>The compensation functions</u>

$$m_f(dG,a) = \frac{1}{2\pi} \int_{dG} \log \frac{1}{\|f;a\|} \, d^\perp \psi \wedge \chi \geqq 0$$

$$m_f^0(dg,a) = \frac{1}{2\pi} \int_{dg} \log \frac{1}{\|f;a\|} \, d^\perp \psi \wedge \chi \geqq 0$$

are defined and non-negative. Also the <u>deficit</u>

$$D_f(G,a) = \frac{1}{2\pi} \int_{G-\overline{g}} \log \frac{1}{\|f,a\|} \, dd^\perp \psi \wedge \chi$$

is defined, but does not have a fixed sign in general.

The characteristic function of $f: M \longrightarrow \mathbb{P}(V)$ is defined by

$$T_f(G) = \int_G \psi f^*(\ddot{\omega}) \wedge \chi \geqq 0$$

Also the integral

$$A_f(G) = \int_G f^*(\ddot{\omega}) \wedge \chi \geqq 0$$

exists.

The First Main Theorem states

$$T_f(G) = N_f(G;a) + m_f(dG;a) - m_f^0(dg;a) - D_f(G;a)$$

Because $m_f(dG;a) \geqq 0$, this implies

$$N_f(G;a) \leqq T_f(G) + m_f^0(dg;a) + D_f(G;a)$$

If g is fixed and if G exhausts M with suitable choice of ψ, then m_f^0 can be estimated, but nothing much can be said without an estimate on the growth of the deficit $D_f(G;a)$.

The freedom in the choice of ψ permits several options. In [36], the form χ was taken positive definite. Then the Dirichlet problem

$$dd^{\perp}\psi \wedge \chi = 0 \qquad \text{on } G - \overline{g}$$

$$\psi | dG = 0$$

$$\psi | dg = R > 0$$

has one and only one solution. The constant $R = R(G)$ is determined such that

$$\frac{1}{2\pi} \int_{dG} d^{\perp}\psi \wedge \chi = \frac{1}{2\pi} \int_{dg} d^{\perp}\psi \wedge \chi = 1$$

Obviously, $g \neq \emptyset$ has to be assumed. The First and Second Main Theorem and the general defect relation were proved in [36]. This theory was applied in [37] to show that a pure m-dimensional analytic subset A of V with $0 < m \leqq n$ is algebraic if and only if its projective volume

$$\int_A \ddot{\omega}^m < +\infty$$

is finite. However all attempts to utilize this approach to solve the Bezout problem of Griffiths for transcendental analytic subsets of V have failed. All information on the complex structure of M is packed into the solution ψ of the Dirichlet problem. In order to obtain asymptotic estimates of $n_f(G,a)$ from estimates of $N_f(G,a)$ if G exhausts M, sharp asymptotic a priory estimates of ψ from below would be needed. This type of global analysis of elliptic differential equations on open complex manifolds is not available at the present time.

Therefore, another approach will be taken, to obtain deficit and Bezout estimates. The function ψ will be defined by an exhaustion. Then g becomes immaterial and can be taken as the empty set.

An <u>exhaustion</u> $\tau: M \to \mathbb{R}$ is a non-negative function of class C^{∞} such that

$$G_r = \{x \in M | \tau(x) < r\}$$

is relative compact in M and such that

$$\Gamma_r = \{x \in M | \tau(x) = r\}$$

has measure zero on M. Let $E(\tau)$ be the set of regular values of τ. If $r \in E(\tau)$, then Γ_r is a C^{∞}-boundary of G_r. Without further notice all boundary integrals over Γ_r appearing in this introduction are taken only if $r \in E(\tau)$. By Sard's theorem, almost all $r > 0$ belong to $E(\tau)$.

The exhaustion τ is assumed to be <u>pseudo-convex</u>, that is, the form

$$\varphi = \frac{1}{4\pi} dd^c\tau \geqq 0$$

is non-negative and positive outside a set of measure zero. Take

$$\chi = \varphi^{m-1}.$$

If $r > 0$, define $\psi_r = \text{Max}\ (0, r-\tau)$. Then $(G_r, \emptyset, \psi_r)$ is a bump for each $r \in E(\tau)$. Hence the theory can be translated. Define

$$A_f(r) = \int_{G_r} f^*(\ddot{\omega}) \wedge \varphi^{m-1}$$

$$T_f(r) = \int_0^r A_f(t)dt = \int_{G_r} \psi_r f^*(\ddot{\omega}) \wedge \varphi^{m-1}$$

$$n_f(r,a) = \int_{\gamma_f(a) \cap G_r} \delta_f^a \ \varphi^{m-1} \geqq 0$$

$$N_f(r,a) = \int_0^r n_f(t;a)dt = \int_{\gamma_f(a)} \psi_r \ \delta_f^a \ \varphi^{m-1} \geqq 0$$

$$m_f(r,a) = \frac{1}{2\pi} \int_{\Gamma_r} \log \frac{1}{\|f,a\|} \ d^c\tau \wedge \varphi^{m-1} \geqq 0$$

$$D_f(r,a) = \frac{1}{2\pi} \int_{G_r} \log \frac{1}{\|f,a\|} \ dd^c\tau \wedge \varphi^{m-1} \geqq 0$$

$$\Phi(r) = \int_{G_r} \varphi^m$$

$$\Phi_1(r) = \int_0^r \Phi(t)dt = \frac{1}{2\pi} \int_{G_r} d\tau \wedge d^c\tau \wedge \varphi^{m-1}$$

The First Main Theorem

$$T_f(r) = N_f(r,a) + m_f(r,a) - D_f(r,a)$$

holds for all $a \in \mathbb{P}(V^*)$ and $r \in E(\tau)$. Since T_f, N_f, D_f are continuous for all $r > 0$, the definition of m_f is extended continuously to all $r > 0$ and the identity hold for all $r > 0$.

The deficit estimate is obtained by the use of the Ahlfors estimates and the Plücker difference formula. In the case $m = 1$, they involve the associated maps defined by the derivatives of f. In the case $m > 1$, there is a crowd of partial derivatives. To select invariant combinations of these partial derivatives at random does not help much.

The intrinsic nature of the one dimensional proof has to be understood, in order to obtain an useful definition of associated maps. A solution to the problem was achieved in [36] by differentiation in the direction of a holomorphic vector field. For technical reasons the vector field is not taken as a section in the holomorphic tangent bundle T of M, but contravariantly in the exterior product $\underset{m-1}{\Lambda} T^*$ of the cotangent bundle. Therefore the vector field is defined by a holomorphic differential form B of bidegree (m-1,0) on M. If

$$\alpha = (\alpha_1, \ldots, \alpha_m) : U_\alpha \to U'_\alpha$$

is a <u>patch</u>, i.e. a biholomorphic map of an open subset U_α of M onto an open subset U'_α of $\mathbb{C}^m$, then B is given in these coordinates by

$$B = \sum_{\nu=1}^{m} B_\nu^\alpha (-1)^\nu d\alpha_1 \wedge \cdots \wedge d\alpha_{\nu-1} \wedge d\alpha_{\nu+1} \wedge \cdots \wedge d\alpha_m$$

If h is a holomorphic vector function on U_α, then

$$h'_\alpha = \sum_{\nu=1}^{m} B_\nu^\alpha \frac{\partial h}{\partial \alpha_\nu}$$

is defined with

$$dh \wedge B = h'_\alpha \, d\alpha_1 \wedge \cdots \wedge d\alpha_m$$

The operator can be iterated and extended to vector functions.

Let $\mathfrak{w} : U_\alpha \to V$ be an irreducible representation of f on U_α. If $0 \leqq p \leqq n$ define

$$\mathfrak{w}_{p\alpha} = \mathfrak{w} \wedge \mathfrak{w}'_\alpha \wedge \mathfrak{w}''_\alpha \wedge \cdots \wedge \mathfrak{w}^{(n)}_\alpha$$

The map f is said to <u>general for B</u> if and only if $\mathfrak{w}_{n\alpha} \not\equiv 0$ for at least one choice of α and $\mathfrak{w}$. Then $(\mathfrak{w}_{p\alpha})^{-1}(0)$ is thin analytic in U_α for all choices of α, $\mathfrak{w}$ and $0 \leqq p \leqq n$. A change of p and α will multiply $\mathfrak{w}_{p\alpha}$ by a nowhere zero holomorphic function. Therefore one and only one meromorphic map

$$f_p : M \longrightarrow \mathbb{P}(\textstyle\bigwedge_{p+1} V)$$

is defined on M such that $f_p = \mathbb{P} \circ \mathfrak{w}_{p\alpha}$ on $U_\alpha - (\mathfrak{w}_{p\alpha})^{-1}(0)$. The meromorphic map f_p is called the <u>pth associated map of f for B</u>. If $m = 1$, then $B = 1$ will produce the classical definition of the associated maps.

The concept of contravariant differentiation defined by B was introduced in [36] and yielded defect relations in several complex variables already 16 years ago. The concept will give the deficit and Bezout estimates here. Differentiation in the direction of a covariant vector field is nothing new and differentiation in the direction of a contravariant vector fields seems to be easier to handle. Holomorphic differential forms abound on Stein manifolds. Of course, there is no unique intrinsic choice in the selection of B, χ respectively the exhaustion τ. However, there is no unique intrinsic choice in the selection of a connection or a Riemannian metric on a differential manifold. Nevertheless differential geometry is build on these concepts. So, the introduction of the forms B and χ, respectively of the exhaustion τ, to measure the growth of f should not be forbidden. After all, there is no length without a yardstick.

Another difficulty may be the assumption that f be general for B. If m = 1 = B, this is equivalent to f(M) not being contained in a hyperplane. Special care has been taken to ascertain a similar result if m > 1. In section 9, the following existence result is obtained.

Assume f: $M \longrightarrow \mathbb{P}(V)$ is a meromorphic map, such that f(M) is not contained in a hyper-plane of $\mathbb{P}(V)$. Assume a holomorphic map $\mathfrak{w}: M \to W$ into a complex vector space W is given such that the restriction $\mathfrak{w}: U \to W$ to some open subset $U \neq \emptyset$ of M provides an immersion. (If M is Stein, an embedding $\mathfrak{w}: M \to \mathbb{C}^{2m+1}$ exists). Then there exists a holomorphic differential form B^0 of bidegree (m-1,0), whose coefficients are polynomials on W of at most degree n-1, such that f is general for $B = \mathfrak{w}^*(B^0)$. If $M \subset V = W$ and if is the inclusion, then the growth of B^0 and hence the growth of B is easily estimated.

Let u be an unbounded function of class C^∞ on the non-negative axis with $u(x) \geqq 0$, $u'(x) > 0$ and $u''(x) \geqq 0$ for all $x \geqq 0$. Then a is said to be a convex change of scale. The function $u \circ \tau$ is again a pseudo-convex exhaustion. Define

$$\varphi_u = \frac{1}{2\pi} dd^c u \circ \tau$$

$$\eta_u(r) = \int_0^r \frac{dt}{(u'(t))^{m-1}}$$

$$\psi_{ur} = \text{Max}\,(0, \eta_u(r) - \eta_u \circ \tau).$$

Then T_f is invariant under the convex change of scale u by the identity

$$T_f(r) = \int_0^r A_f(t) \frac{dt}{(u'(t))^{m-1}} = \int_{G_r} \psi_{ur} f^*(\ddot{\omega}) \wedge \varphi_u^{m-1}$$

Hence $T_f(r)$ is the characteristic for the bump $(G_r, \emptyset, \psi_{ur})$ with $\chi = \varphi_u^{m-1}$ for every such convex function u. Similar invariance properties hold for N_f, m_f, D_f, Φ_1. In the classical theory the differential dt in the definition of T_f is replaced by $\frac{dt}{t}$ or sometimes Chern [7] and [39] by $\frac{dt}{t^k}$. This is due to a change of scale. The favored choices are those, which are either adapted to the problem at hand, or which eliminate the deficity by

$$dd^{\perp} \psi_{ru} \wedge \varphi_u^{m-1} = 0$$

In the case M = V, this is possible for $\varphi_u = \omega$ and $\varphi_u = \upsilon$.

Part of the investigations will be carried out on pseudo-convex space. Later, a Stein manifold will be needed. So, for the introduction, assume that M is Stein. Assume that a convex change of scale u is given such that $\varphi_u > 0$ on M. Define

$$i_0 = \left(\frac{i}{2\pi}\right)^{m-1} (-1)^{\frac{(m-1)(m-2)}{2}}$$

Take $r \geqq 0$. If $G_r \neq \emptyset$, a constant $C \geqq 1$ exists such that

$$i_0 B \wedge \overline{B} \leqq (C\varphi_u)^{m-1}$$

on $G_r \cup \Gamma_r$. Let $Y_u(r)$ be the infimum of all these constants C. If $G_r = \emptyset$, define $Y_u(r) = 1$. The function Y_u increases with

$$i_0 B \wedge \overline{B} \leqq [(Y_u \circ \tau)\, \varphi_u]^{m-1}$$

on M. Then $Z_u = Y_u u'$ measures the growth of B.

The set I_{f_p} of indetermancies of f_p is analytic in M with $\dim I_{f_p} \leqq m-2$. Define

$$H_p = i_0 B \wedge \overline{B} \wedge f_p^*(\ddot{\omega}_p) \geqq 0$$

on $M - I_{f_p}$. A function $h_{up} \geqq 0$ of class C^∞ on $M - I_{f_p}$ is uniquely defined by

$$H_p = h_{up}\, \varphi_u^m$$

The functions h_{up} for $p = 0,1,\ldots,n-1$ measure the inflection of f. Their geometric meaning is extensively investigated in section 13. The <u>inflection functions</u> of f are defined by

$$s_f^p(r;u) = \frac{1}{2} \int_{G_r} \log \frac{1}{h_{up}}\, \varphi^m$$

$$\overset{+}{S}{}^p_f(r;u) = \frac{1}{2}\int_{G_r} \log^+ \frac{1}{h_{up}}\,\varphi^m \geqq 0$$

$$\overset{+}{S}_f(r;u) = \sum_{p=0}^{n-1} \overset{+}{S}{}^p_f(r;u) \geqq 0$$

The examples of Cornalba-Shiffman [9] and Griffiths [16] show, that they cannot be majorized by T_f in general.

Another inflection measure is provided by the <u>p^{th} stationary divisor v_p</u>. Let $\alpha\colon U_\alpha \to U'_\alpha$ be a patch on M and let $\mathfrak{w}\colon U_\alpha \to V$ be an irreducible representation on M. Assume that U_α is a Cousin II domain. Then a holomorphic function $g_p\colon U_\alpha \to \mathbb{C}$ and an irreducible representation $\mathfrak{v}_p$ of f_p on U_α exist such that $\mathfrak{w}_{p\alpha} = g_p\,\mathfrak{v}_p$. Let $\delta_{p\alpha}$ be the zero divisor of g_p on U_α. Then

$$v_p = \delta_{p-1,\alpha} - 2\delta_{p\alpha} + \delta_{p+1,\alpha} \geqq 0$$

does not depend on the choices of α and $\mathfrak{w}$. It defines $v_p \geqq 0$ as a divisor on M.

Let Ric_u be the Ricci form of the volume form $\varphi_u^m > 0$ on M. The Ricci function of M is defined by

$$Ric_u(r) = \int_0^r \int_{G_t} Ric_u \wedge \varphi^{m-1}\,dt$$

$$Ric_u^+(r) = \int_0^r \int_{G_t} \text{Max}\,(0, Ric_u \wedge \varphi^{m-1})dt \geqq 0$$

A function $\rho_u \geqq 0$ of class C^∞ on M is defined by

$$\rho_u\varphi_u^m = \frac{1}{2\pi}\,d\tau \wedge d^c\tau \wedge \varphi^{m-1}$$

Define

$$Q_u(r) = \int_{G_r} \log^+ \rho_u \, \varphi^m \geqq 0$$

Given $z \in \mathbb{P}(\underset{p+1}{\Lambda} V)$ and $a \in \mathbb{P}(V^*)$, take any $\mathfrak{z} \in \mathbb{P}^{-1}(z)$ and $\alpha \in \mathbb{P}^{-1}(a)$. Then

$$\| z;a \| = \frac{|\mathfrak{z} \llcorner \alpha|}{|\mathfrak{z}| \, |\alpha|}$$

is well defined and agrees with the previous definition if $p = 0$.

The deficit estimates are based on the following two results.

1. Ahlfors estimate. (Theorem 10.11)

A constant $c_n > 2$, depending on n only, exists such that for all $a \in \mathbb{P}(V^*)$, all $r \in \mathbb{R}$ with $r > 0$, all $\lambda \in \mathbb{R}$ with $0 < \lambda < 1$ and all $p \in \mathbb{Z}$ with $0 \leqq p < n$, the following estimate holds

$$\int_0^r \Big[\int_{G_t} \frac{\| f_{p+1};a \|}{\| f_p;a \|^{2\lambda}} H_p \Big] \frac{dt}{(Z_u(t))^{m-1}} \leqq \frac{c_n \, e^p}{(1-\lambda)^2} \left[T_{f_p}(r) + \Phi(r) \right]$$

2. Plückers Difference Formula. (Theorem 12.1)

$$N_{v_p}(r) + T_{f_{p-1}}(r) - 2T_{f_p}(r) + T_{f_{p+1}}(r) =$$

$$= \Omega_f^p(r,u) + S_f^p(r;u) + \mathrm{Ric}_u(r)$$

Uniform Deficit Estimate. (Theorem 12.13)

A constant $C > 0$, depending on n, m and $u'(0)$ only, exists such that for all $a \in \mathbb{P}(V^*)$ and all numbers θ, r with $1 < \theta \leqq 2 < r$ the following estimate holds

$$D_f(r;a) \leqq 2^{2n+1}\,\Phi(r)\,[\log^+ T_f(\theta r) + \log^+ \Phi(\theta r) + \log^+ Q_u(\theta r) + (m-1)\log^+ Z_u(\theta r) + \log^+ \overset{+}{S}_f(\theta r, u) + \log \frac{\theta}{\theta - 1} + C] + 2^n\,[1 + \overset{+}{S}_f(r;u)]$$

Observe that this estimate is absolutely uniform. The constants do not depend on θ, r, a, f. Observe that only Z_u on $\overset{+}{S}_f$ depend on B. A similar estimate is obtained for $\theta = 1$, but then the estimate holds only outside a set of measure zero, which may depend on a.

A meromorphic map is said to be steady if $h_{up} \geqq c > 0$ for a constant c. This radical assumptions was introduced by Griffiths [16]. If f is steady then

$$\overset{+}{S}_f(r;u) \leqq c_0\,\Phi(r)$$

for some constant $c > 0$. The deficit estimate simplifies (Theorem 13.4 and Theorem 13.5). The meaning of "steady" is explored in section 13.

The map f grows sufficiently if and only of constants c_ν for $\nu = 1,2,3,4$ and $0 \leqq \varepsilon < \frac{1}{2}$ exist such that

$$\Phi(r) \leqq c_1\, T_f(r)^{\varepsilon}$$

$$\log^+ Z_u(r) \leqq c_2\, T_f(r)^{\varepsilon}$$

$$\log^+ Q_u(r) \leqq c_3\, T_f(r)^{\varepsilon}$$

$$\log^+ \mathrm{Ric}_u^+(r) \leqq c_4\, T_f(r)^{\varepsilon}$$

for all sufficiently large r. The map f _grows slowly_ if a constant $c > 0$ exists such that $T_f(\theta r) \leqq c\, T_f(r)$ for all sufficiently large r. The defect $\delta_f(a)$ of f for $a \in \mathbb{P}(V^*)$ is defined by

$$\delta_f(a) = \liminf_{r \to +\infty} \frac{m_f(r,a)}{T_f(r)} \geqq 0$$

If the steady meromorphic map grows sufficiently, then

$$0 \leqq \delta_f(a) \leqq 1$$

with $\delta_f(a) = 1$ if $f^{-1}(\ddot{E}[a]) = \emptyset$. If the steady map grows slowly but sufficiently, then

$$\frac{D_f(r;a)}{T_f(r)} \Longrightarrow 0 \qquad \text{for } r \to \infty$$

uniformly for all $a \in \mathbb{P}(V^*)$ and

$$\delta_f(a) = 1 - \limsup_{r \to +\infty} \frac{N_f(r;a)}{T_f(r)} .$$

These results can be applied to obtain Bezout estimate. Again, let V be a hermitian vector space of dimension n+1 with $n > 0$. If $A \subseteq V$, define

$$A[r] = \{ \mathfrak{z} \in A \mid |\mathfrak{z}| \leq r\}$$

$$A(r) = \{ \mathfrak{z} \in A \mid |\mathfrak{z}| < r\}$$

$$A\langle r\rangle = \{ \mathfrak{z} \in A \mid |\mathfrak{z}| = r\}$$

If A is an analytic subset of pure dimension p of V, define

$$n_A(r) = \# A[r] \qquad \text{if } p = 0$$

$$n_A(r) = \frac{1}{r^{2p}} \int_{A[r]} \upsilon^p = \int_{A[r]} \omega^p + n_A(0) \quad \text{if } p > 0$$

Here $n_A \geqq 0$ increases with $n_A(0) = \lim_{r \to 0} n_A(r)$. The Lelong number $n_A(0)$ is an integer by Thie [46]. If $0 \notin A$, define

$$N_A(r) = \int_0^r n_A(t) \frac{dt}{t}$$

Let M be a connected, smooth, closed, complex submanifold of dimension $m > 0$ of V with $M[1] = \emptyset$. Let $\mathfrak{10} : M \to V$ be the inclusion. Assume that M is not contained in any linear subspace E[a] of V. Then $M_a = M \cap E[a]$ is thin analytic in M.

Given $a \in \mathbb{P}(V^*)$, take any $\alpha \in \mathbb{P}^{-1}(a)$. Then the zero divisor $\delta_M^a = \delta_\alpha \circ \mathfrak{10}$ on M does not depend on the choice of α.

Define

$$n_M(r;a) = \sum_{\mathfrak{z}\in M_a[r]} \delta_M^a(\mathfrak{z}) \qquad \text{if } m = 1$$

$$n_M(r,a) = \frac{1}{r^{2n-2}} \int_{M_a[r]} \delta_M^a \, \upsilon^{m-1}$$

$$\qquad\qquad\qquad\qquad\qquad\qquad \text{if } m > 1$$

$$= \int_{M_a[r]} \delta_M^a \, \omega^{m-1}$$

$$N_M(r,a) = \int_0^r n_M(t;a) \frac{dt}{t}$$

The Bezout problem seeks to estimate $N_M(r,a)$ (or $n_M(r;a)$) uniformly in terms of $n_M(r)$ and $N_M(r)$ and perhaps other invariants of M.

Now, the previous results on value distribution are applied to the holomorphic map

$$f = \mathbb{P} \circ \mathfrak{w} : M \to \mathbb{P}(V).$$

Define $\sigma = \frac{1}{4\pi} d^c \log \tau_0 \wedge \omega^{m-1}$. Then

$$m_M(r;a) = \frac{1}{2\pi} \int_{M\langle r\rangle} \log \frac{1}{\| f;a \|} \sigma$$

$$D_M(r;a) = \frac{1}{2\pi} \int_{M(r)} \log \frac{1}{\| f;a \|} \omega^m$$

Then

$$N_M(r) = N_M(r;a) + m_M(r;a) - D_M(r;a)$$

by the First Main Theorem.

If $0 \leqq p < n$, define the Grassmann cone by

$$\tilde{G}_p(V) = \{ \mathfrak{z}_0 \wedge \cdots \wedge \mathfrak{z}_p \mid \mathfrak{z}_\mu \in V\}$$

The Grassmann manifold of (p+1)-dimensional linear subspaces of V is given by

$$G_p(V) = \mathbb{P}(\tilde{G}_p(V)) \subseteq \mathbb{P}(\textstyle\bigwedge_{p+1} V)$$

and is a smooth, compact, connected, complex submanifold of dimension $(p+1)\cdot(n-p)$ of $\mathbb{P}(\bigwedge_{p+1} V)$.

Take $\mathfrak{z} \in M$. Then $\tilde{\Delta}(\mathfrak{z}) \in \tilde{G}_{m-1}(V)$ exist such that

$$\{ \mathfrak{z} + \mathfrak{y} \mid \mathfrak{y} \wedge \tilde{\Delta}(\mathfrak{z}) = 0\}$$

is the tangent plane of M at $\mathfrak{z}$ Then $\Delta(\mathfrak{z}) = \mathbb{P}(\tilde{\Delta}(\mathfrak{z}))$ is uniquely defined. The Gauss map

$$\Delta\colon M \to G_{m-1}(V) \subseteq \mathbb{P}(\textstyle\bigwedge_m V)$$

is holomorphic. Let Ric_M be the Ricci form of $\mathfrak{io}^*(\upsilon^m)$. Then

$$\mathrm{Ric}_M = \Delta^*(\ddot{\omega}_{m-1}) \geqq 0$$

Hence the characteristic function of the Gauss map is the Ricci function of M

$$T_\Delta(r) = \int_0^r \int_{M[t]} \Delta^*(\ddot{\omega}_{m-1}) \wedge \omega^{m-1} \frac{dt}{t}$$

Take a holomorphic differential form B^0 of bidegree $(m-1,0)$ such that B^0 has polynomial coefficients of atmost degree $n-1$ and such that $f = P \circ \psi$ is general for $B = \psi^*(B^0)$. As seen before, such a form exists. If $m = 1$, take $B^0 = 1$. The associated maps f_p of $f = P \circ \psi$ are defined. Also the differential forms

$$H_p = i_0 B \wedge \overline{B} \wedge f_p^*(\ddot{\omega}_p)$$

are almost everywhere defined. Again h_{up} is defined by

$$H_p = h_{up}\ \psi^*(\upsilon^m)$$

The <u>inflection functions</u>

$$\overset{+}{s}{}^p_M(r) = \frac{1}{2} \int_{M(r)} \log^+ \frac{1}{h_{up}}\ \omega^m$$

$$\overset{+}{s}_M(r) = \sum_{p=0}^{n-1} \overset{+}{s}{}^p_M(r)$$

are defined.

The manifold M is said to be <u>steady</u> if a constant $c > 0$ exists such that $h_{up} \geqq c > 0$ almost everywhere on M for $p = 0,1,\ldots,n-1$.

There exists a number $r_1 > 1$ such that $M[r] = \emptyset$ if $r < r_1$ and $M[r] \neq \emptyset$ if $r \geqq r_1$.

<u>Uniform Bezout Estimate</u>. (Corollary 14.11)

Take $0 < \varepsilon < 1$. Then a constant $c > 0$ exists such that for all $a \in \mathbb{P}(V^*)$ and all $r > r_1 + e^2$ the following estimate holds

$$N_M(r;a) \leqq c\, n_M(r)\, [\log^+ N_M(r+\varepsilon) + \log^+ T_\Delta(r+\varepsilon) + \log r]$$

$$+ 2^{2n+1}\, n_M(r) \log^+ \overset{+}{s}_M(r+\varepsilon) + 2^n\, \overset{+}{s}_M(r)$$

If in addition M is steady for B, then

$$N_M(r;a) \leqq c_1\, n_M(r)\, [\log^+ N_M(r+\varepsilon) + \log^+ T_\Delta(r+\varepsilon) + \log r]$$

for some constant c_1 and all $a \in \mathbb{P}(V^*)$ and all $r > r_1 + e^2$.

These deficit and Bezout estimates may look weak, because not too much is known about the remainder terms. However, they are at least uniform estimates of some type. The explicit exposure of the remainder terms may lead to better results.

II. GRASSMANN ALGEBRA IN HERMITIAN VECTOR SPACES

In this section, some well known facts on Grassmann algebra will be stated to ascertain the notations.

If A is a partially ordered set, denote

$$A[s,r] = \{x \in A \mid s \leqq x \leqq r\}$$

$$A(s,r] = \{x \in A \mid s < x \leqq r\}$$

$$A[s,r) = \{x \in A \mid s \leqq x < r\}$$

$$A(s,r) = \{x \in A \mid s < x < r\}$$

Here s and r may not be elements of A; only the notations $s < x < r$ etc. must make sense.

For $p \in \mathbf{Z}[0,n]$, define $\mathfrak{M}(p,n)$ as the set of all maps $\mu: \mathbf{Z}[0,p] \to \mathbf{Z}[0,n]$. Let $\mathfrak{J}(p,n)$ be the subset of all increasing, injective maps $\mu \in \mathfrak{M}(p,n)$. The cardinality of $\mathfrak{J}(p,n)$ is

$$\# \mathfrak{J}(p,n) = \binom{n+1}{p+1}$$

The set of all bijective maps $\mu \in \mathfrak{M}(n,n)$ is the symmetric group $\mathfrak{S}(n)$. The homomorphism sign: $\mathfrak{S}(n) \to Z$ is defined. Define sign $\mu = 0$ if $\mu \in \mathfrak{M}(n,n) - \mathfrak{S}(n)$.

For each $\mu \in \mathfrak{J}(p,n)$, one and only one $\mu^{\perp} \in \mathfrak{J}(n-p-1,n)$ exists such that the image sets of μ and $\mu^{\perp}$ are disjoint. A permutation $\hat{\mu} \in \mathfrak{S}(n)$ is defined by $\hat{\mu}(x) = x$ if $x \in \mathbf{Z}[0,p]$ and $\hat{\mu}(x) = \mu^{\perp}(x-p-1)$ if $x \in \mathbf{Z}[p+1,n]$. Define sign $\mu = \text{sign } \hat{\mu}$. Obviously $(\mu^{\perp})^{\perp} = \mu$ and

$$\text{sign } \mu^{\perp} = (-1)^{(p+1)(n-p)} \text{ sign } \mu$$

The <u>Kronecker symbol</u> of a set S is a map

$$\delta: S \times S \to \mathbf{Z}[0,1]$$

such that

$$\delta_{xy} = \begin{cases} 1 & \text{if } x = y \\ 0 & \text{if } x \neq y \end{cases}$$

Let V be a complex vector space of pure dimension n+1 with $n \geqq 0$. The p-folded exterior product $\underset{p}{\Lambda} V$, the tensor product $\underset{p}{\otimes} V$ and the direct sum $\underset{p}{\oplus} V$ are defined.

If $\mathfrak{w}_0, \ldots, \mathfrak{w}_q$ are elements of V and if $\mu \in \mathfrak{T}(p,q)$, define

$$\mathfrak{w}_\mu = \mathfrak{w}_{\mu(0)} \wedge \cdots \wedge \mathfrak{w}_{\mu(p)}$$

If $\mathfrak{w} = (\mathfrak{w}_0, \ldots, \mathfrak{w}_n)$ is a base of V, then

$$\mathfrak{w}^p = \{ \mathfrak{w}_\mu \mid \mu \in \mathfrak{T}(p,n) \}$$

is a base of $\underset{p+1}{\Lambda} V$.

The <u>Grassmann cone</u> of order p is defined by

$$\tilde{G}_p(V) = \{ \mathfrak{w}_0 \wedge \cdots \wedge \mathfrak{w}_p \mid \mathfrak{w}_\mu \in V \}.$$

The <u>dual vector space</u> V* of V consists of all linear functions $\alpha: V \to \mathbb{C}$. The <u>inner product</u> between $\mathfrak{z} \in V$ and $\alpha \in V^*$ is defined by

$$(\mathfrak{z}; \alpha) = \mathfrak{z} \llcorner \alpha = \alpha(\mathfrak{z})$$

For every base $\mathfrak{w} = (\mathfrak{w}_0, \ldots, \mathfrak{w}_n)$ of V, one and only one base $\varepsilon = (\varepsilon_0, \ldots, \varepsilon_n)$ of V^*, called the <u>dual base</u> exists such that

$$(\mathfrak{v}_\mu, \varepsilon_\nu) = \delta_{\mu\nu}$$

for $\mu, \nu = 0, \ldots, n$. The identification $(V^*)^* = V$ is made such that

$$\mathfrak{x}(\alpha) = (\mathfrak{x}; \alpha) = \alpha(\mathfrak{x})$$

for all $\mathfrak{x} \in V$ and $\alpha \in V^*$. Then

$$(\underset{p}{\Lambda} V^*) = (\underset{p}{\Lambda} V)^*$$

are identified by

$$(\mathfrak{x}_1 \wedge \cdots \wedge \mathfrak{x}_p; \alpha_1 \wedge \cdots \wedge \alpha_p) = \det(\mathfrak{x}_\mu; \alpha_\nu)$$

This identification is consistent with the exterior algebra. If ε is the dual base of the base $\mathfrak{v}$ of V, then ε^p is the dual base of the base $\mathfrak{v}^p$ of $\underset{p+1}{\Lambda} V$.

Let $\sigma: V \to W$ be a linear map between the complex vector spaces V and W. It induces linear maps $\sigma^*: W^* \to V^*$ and $\sigma_p: \underset{p}{\Lambda} V \to \underset{p}{\Lambda} V$ such that

$$\sigma^*(\alpha) = \alpha \circ \sigma \qquad \text{if } \alpha \in W^*$$

$$\sigma_p(\mathfrak{x}_1 \wedge \cdots \wedge \mathfrak{x}_p) = \sigma(\mathfrak{x}_1) \wedge \cdots \wedge \sigma(\mathfrak{x}_p) \quad \text{if } \mathfrak{x}_\mu \in V$$

Then

$$\sigma_{p+q}(\mathfrak{x} \wedge \mathfrak{y}) = \sigma_r(\mathfrak{x}) \wedge \sigma_q(\mathfrak{y})$$

if $\mathfrak{x} \in \underset{p}{\Lambda}V$ and $\mathfrak{y} \in \underset{q}{\Lambda}V$. Abbreviate $\sigma = \sigma_p$ for all p. Observe

$$(\sigma^*)_p = (\sigma_p)^*$$

An additive map $\sigma\colon V \to W$ is said to be antilinear if $\sigma(a\mathfrak{x}) = \bar{a}\,\sigma(\mathfrak{x})$ for all $\mathfrak{x} \in V$ and all $a \in \mathbb{C}$. Similar properties hold for antilinear maps.

Let $\mathfrak{n} = (\mathfrak{n}_0, \ldots, \mathfrak{n}_n)$ be a base of V. For each $p \in \mathbb{Z}[0,n]$ a linear isomorphism

$$D_{\mathfrak{n}} : \underset{p}{\Lambda}V \to \underset{n+1-p}{\Lambda} V^*$$

is defined by

$$(\mathfrak{x}; D_{\mathfrak{n}}\mathfrak{w})\, \mathfrak{n}_0 \wedge \cdots \wedge \mathfrak{n}_n = \mathfrak{x} \wedge \mathfrak{w}$$

for all $\mathfrak{x} \in \underset{n+1-p}{\Lambda} V$ if $\mathfrak{w} \in \underset{p}{\Lambda}V$. Then

$$D_{\mathfrak{n}} : \tilde{G}_{p-1}(V) \to \tilde{G}_{n-p}(V).$$

Let $\varepsilon = (\varepsilon_0, \ldots, \varepsilon_n)$ be the dual base to $\mathfrak{n}$. If $\mathfrak{w} \in \underset{p}{\Lambda}V$, then

$$D_\varepsilon \circ D_{\mathfrak{n}}(\mathfrak{w}) = (-1)^{p(n+1-p)}$$

The linear map $D_{\mathfrak{n}} : V \to V^*$ defines the dual map $D^*_{\mathfrak{n}}: V^{**} \to V^*$. Here $V^{**} = V$. Then $D^*_{\mathfrak{n}} = (-1)^{p(n+1-p)} D_{\mathfrak{n}}$ on $\underset{n+1-p}{\Lambda} V$.

Let p and q be integers with $1 \leqq q \leqq p \leqq n+1$. The <u>interior product</u> $\mathfrak{x} \llcorner \alpha \in \underset{p-q}{\Lambda} V$ of $\mathfrak{x} \in \underset{p}{\Lambda} V$ with $\alpha \in \underset{q}{\Lambda} V^*$ is defined by

$$(\mathfrak{x} \llcorner \alpha;\beta) = (\mathfrak{x};\alpha \wedge \beta)$$

for all $\beta \in \underset{p-q}{\Lambda} V^*$. If $p = q$, then $\mathfrak{x} \llcorner \alpha = (\mathfrak{x};\alpha)$. If $\mathfrak{x} \in \underset{p}{\Lambda} V$, if $\alpha \in \underset{q}{\Lambda} V^*$ and if $\beta \in \underset{r}{\Lambda} V^*$ with $p \geqq q+r$, then

$$(\mathfrak{x} \llcorner \alpha) \llcorner \beta = \mathfrak{x} \llcorner \alpha \wedge \beta$$

If ε is a base of V^*, if $\alpha \in \underset{p}{\Lambda} V^*$ and if $\beta \in \underset{q}{\Lambda} V^*$ with $p+q \leqq n+1$, then

$$D_{\varepsilon}(\alpha \wedge \beta) = (D_{\varepsilon}\alpha) \llcorner \beta$$

Let $p \in \mathbb{Z}[1,n+1]$. Take $\mathfrak{x}_{\mu} \in V$ for $\mu = 0,1,\ldots,p+1$ and $\alpha_{\mu} \in V^*$ for $\mu = 0,\ldots,p-1$. Define

$$\alpha = \alpha_0 \wedge \cdots \wedge \alpha_{p-1}$$

$$\mathfrak{y}_q = \mathfrak{x}_0 \wedge \cdots \wedge \mathfrak{x}_q \quad \text{for } 0 \leqq q \leqq p+1$$

Then the following <u>transportation</u> formula holds

$$(2.1) \qquad (\mathfrak{y}_p \llcorner \alpha) \wedge [(\mathfrak{y}_{p-1} \wedge \mathfrak{x}_{p+1}) \llcorner \alpha] = (\mathfrak{y}_{p-1};\alpha)\, \mathfrak{y}_{p+1} \llcorner \alpha$$

A positive definite hermitian form h on V is called a <u>hermitian product</u> on V and denoted by

$$h(\mathfrak{x},\mathfrak{y}) = (\mathfrak{x}|\mathfrak{y})$$

for all $\mathfrak{x} \in V$ and $\mathfrak{y} \in V$. Then $|\mathfrak{x}| = (\mathfrak{x}|\mathfrak{x})^{\frac{1}{2}}$ is the absolute value of $\mathfrak{x}$. A hermitian product on V induces hermitian products on $\underset{p}{\Lambda}V$ and V^*. An antilinear isomorphism $\mu\colon V \to V^*$ is defined by $\mu(\mathfrak{y})(\mathfrak{x}) = (\mathfrak{x}|\mathfrak{y})$ for all $\mathfrak{x} \in V$ if $\mathfrak{y} \in V$. Then

$$(\mathfrak{x}_1 \wedge \cdots \wedge \mathfrak{x}_p | \mathfrak{y}_1 \wedge \cdots \wedge \mathfrak{y}_p) = \det\,(\mathfrak{x}_\mu|\mathfrak{y}_\nu)$$

$$(\alpha|\mu\mathfrak{x}) = \alpha(\mathfrak{x})$$

for all $\mathfrak{x}_\mu \in V$ and $\mathfrak{y}_\mu \in V$ and all $\alpha \in V^*$ and $\mathfrak{x} \in V$. The antilinear map $\mu\colon V \to V^*$ induces an antilinear map $\mu\colon \underset{p}{\Lambda}V \to \underset{p}{\Lambda}V^*$. If $\alpha \in \underset{p}{\Lambda}V^*$ and $\mathfrak{x} \in \underset{p}{\Lambda}V$, then

$$(\alpha|\mu\mathfrak{x}) = \alpha(\mathfrak{x}).$$

A complex vector space together with an assigned hermitian product is also called a <u>hermitian vector space</u>.

A base $\mathfrak{w} = (\mathfrak{w}_0,\ldots,\mathfrak{w}_n)$ of the hermitian vector space V is said to be <u>orthonormal</u> if $(\mathfrak{w}_\mu|\mathfrak{w}_\nu) = \delta_{\mu\nu}$ for $\mu,\nu = 0,1,\ldots,n$. The dual base ε of an orthonormal base is orthonormal, so is $\mathfrak{w}^p$. If $\mathfrak{w}$ is orthonormal, then

$$(D_{\mathfrak{w}}\mathfrak{x} \mid D_{\mathfrak{w}}\mathfrak{y}) = (\mathfrak{x} \mid \mathfrak{y})$$

for all $\mathfrak{x}, \mathfrak{y} \in V$. If $\mathfrak{x}_0, \ldots, \mathfrak{x}_p$ are vectors in V with $0 \leqq p \leqq n$, an orthonormal base $\mathfrak{w} = (\mathfrak{w}_0, \ldots, \mathfrak{w}_n)$ exists such that

$$\mathfrak{x}_\mu = \sum_{\nu=0}^{\mu} x_{\mu\nu}\, \mathfrak{w}_\nu .$$

The set of all linear isomorphisms $\sigma: V \to V$ is a group GL(V), which is called the <u>general linear group</u> of V. An isomorphism $\sigma \in GL(V)$ is called an isometry if and only if $|\sigma(\mathfrak{x})| = |\mathfrak{x}|$ for all $\mathfrak{x} \in V$. The group U(V) of all isometries is called the <u>unitary group</u> of V. If $\sigma \in U(V)$, then $(\sigma(\mathfrak{x}) | \sigma(\mathfrak{y})) = (\mathfrak{x} | \mathfrak{y})$ for all $\mathfrak{x}, \mathfrak{y} \in V$.

If $A \subseteq V$, define

$$A[r] = \{ \mathfrak{z} \in A \mid |\mathfrak{z}| \leqq r \}$$

$$A(r) = \{ \mathfrak{z} \in A \mid |\mathfrak{z}| < r \}$$

$$A\langle r\rangle = \{ \mathfrak{z} \in A \mid |\mathfrak{z}| = r \}$$

Let $\mathfrak{x}_0, \ldots, \mathfrak{x}_{p+1}$ be vectors in V with $0 < p < n$. Define

$$\mathfrak{w} = \mathfrak{x}_0 \wedge \cdots \wedge \mathfrak{x}_p \in \textstyle\bigwedge_{p+1} V$$

$$\mathfrak{y} = \mathfrak{x}_0 \wedge \cdots \wedge \mathfrak{x}_{p-1} \wedge \mathfrak{x}_{p+1} \in \textstyle\bigwedge_{p+1} V$$

Denote the exterior product on the vector space $W = \underset{p+1}{\Lambda} V$ by $\hat{\wedge}$. Then $\mathfrak{w} \hat{\wedge} \mathfrak{y} \in \underset{2}{\Lambda} W$. Observe that $\underset{2}{\Lambda} W$ is again a hermitian vector space. Then

$$|\mathfrak{x}_0 \wedge \cdots \wedge \mathfrak{x}_{p-1}| \, |\mathfrak{x}_0 \wedge \cdots \wedge \mathfrak{x}_{p+1}| = |\mathfrak{w} \hat{\wedge} \mathfrak{y}| \qquad (2.2)$$

The distance between $0 \neq \mathfrak{x} \in \underset{p+1}{\Lambda} V$ and $0 \neq \mathfrak{y} \in \underset{q+1}{\Lambda} V$ is defined by

$$\|\mathfrak{x} : \mathfrak{y}\| = \frac{|\mathfrak{x} \wedge \mathfrak{y}|}{|\mathfrak{x}|\,|\mathfrak{y}|} = \|\mathfrak{y} : \mathfrak{x}\|$$

If $\mathfrak{y} \in \tilde{G}_p(V)$, then $\|\mathfrak{x} : \mathfrak{y}\| \leqq 1$. The <u>distance</u> between $0 \neq \mathfrak{x} \in \underset{p+1}{\Lambda} V$ and $0 \neq \alpha \in \underset{q+1}{\Lambda} V^*$ with $0 \leqq q \leqq p$ is defined by

$$\|\mathfrak{x} ; \alpha\| = \frac{|\mathfrak{x} \llcorner \alpha|}{|\mathfrak{x}|\,|\alpha|}$$

If $\alpha \in \tilde{G}_q(V)$, then $\|\mathfrak{x} ; \alpha\| \leqq 1$. If $\mathfrak{r}$ is an orthonormal base of V, if $0 \neq \mathfrak{x} \in \underset{p+1}{\Lambda} V$ and $0 \neq \alpha \in \underset{q+1}{\Lambda} V^*$ with $0 \leqq q \leqq p$, then

$$\|D_{\mathfrak{r}} \mathfrak{x} ; \alpha\| = \|\mathfrak{x} ; \alpha\|$$

If $\mu: V \to V^*$ is the antilinear map defined by $\mu(\mathfrak{x})(\mathfrak{y}) = (\mathfrak{y} \mid \mathfrak{x})$ then

$$\|\mathfrak{x} : \mathfrak{v}\|^2 + \|\mathfrak{x} ; \mu \mathfrak{v}\|^2 = 1$$

for all $0 \neq \mathfrak{x} \in \tilde{G}_p(V)$ and all $0 \neq \mathfrak{v} \in V$.

The space $\mathbb{C}^n$ is always considered to be a hermitian vector space with the hermitian product

$$(\mathfrak{x} | \mathfrak{y}) = \sum_{\mu=1}^{n} x_\mu \bar{y}_\mu$$

where $\mathfrak{x} = (x_1, \ldots, x_n)$ and $\mathfrak{y} = (y_1, \ldots, y_n)$.

On $V - \{0\}$, an equivalence relation is defined, such that the equivalence class $\mathbb{P}(\mathfrak{x})$ of $0 \neq \mathfrak{x} \in V$ is defined by

$$\mathbb{P}(\mathfrak{x}) = \{z\mathfrak{x} | 0 \neq z \in \mathbb{C}\}.$$

If $A \subseteq V$, define

$$\mathbb{P}(A) = \{\mathbb{P}(\mathfrak{x}) | 0 \neq \mathfrak{x} \in A\}$$

Then $\mathbb{P}(V)$ is a connected, compact, n-dimensional, complex manifold called the <u>complex projective space</u> of V. Denote $\mathbb{P}_n = \mathbb{P}(C^{n+1})$. The map

$$\mathbb{P}: V - \{0\} \to \mathbb{P}(V)$$

is holomorphic, regular, surjective and 1-fibering. If $p \in \mathbb{Z}[0,n]$, then

$$G_p(V) = \mathbb{P}(\tilde{G}_p(V))$$

is a connected, smooth, compact, complex submanifold of $\mathbb{P}(\underset{p+1}{\Lambda} V)$ called the <u>Grassmann manifold</u> of projective p-planes in $\mathbb{P}(V)$. The dimension of $G_p(V)$ is

$$\dim G_p(V) = (p+1)(n-p)$$

Here $\mathbb{P}(V) = G_0(V)$. The space $G_n(V)$ is a point. If $0 \neq \mathfrak{x} \in \tilde{G}_p(V)$, then

$$E(\mathfrak{x}) = \{ \mathfrak{z} \in V \mid \mathfrak{z} \wedge \mathfrak{x} = 0 \}$$

is a linear subspace of dimension p+1 of V. If $\mathfrak{x} = \mathfrak{x}_0 \wedge \cdots \wedge \mathfrak{x}_p$, then

$$E(\mathfrak{x}) = \mathbb{C}\mathfrak{x}_0 + \ldots + \mathbb{C}\mathfrak{x}_p$$

The projective space $\mathbb{P}(E(\mathfrak{x})) = \ddot{E}(\mathfrak{x})$ of $E(\mathfrak{x})$ is a smooth, connected, compact, p-dimensional, complex submanifold of $\mathbb{P}(V)$. It is called a <u>projective plane of dimension p</u>. If $a \in G_p(V)$, take $\mathfrak{x} \in \mathbb{P}^{-1}(a)$. Then $E(a) = E(\mathfrak{x})$ and $\ddot{E}(a) = \ddot{E}(\mathfrak{x}) = \mathbb{P}(E(a))$ are defined independently of the choice of $\mathfrak{x}$.

If $0 \neq \alpha \in V^*$, then $E[\alpha] = \ker \alpha = \alpha^{-1}(0)$ is a linear subspace of dimension n of V. Also $\ddot{E}[\alpha] = \mathbb{P}(E[\alpha])$ is a projective (n-1)-plane, called a <u>hyperplane</u>. If $a \in \mathbb{P}(V^*)$, take any $\alpha \in \mathbb{P}^{-1}(a)$. Then $E[a] = E[\alpha]$ and $\ddot{E}[a] = \ddot{E}[\alpha] = \mathbb{P}(E[a])$ are defined independently of the choice of α.

Let p and q be integers with $0 \leqq p \leqq q \leqq n-1$. Then

$$F_{pq} = \{(x,y) \in G_p(V) \times G_q(V) \mid \ddot{E}(x) \subseteq \ddot{E}(y)\}$$

is a connected, compact, smooth, complex submanifold of $G_p(V) \times G_q(V)$. The projections $\tau\colon F_{pq} \to G_p(V)$ and $\pi\colon F_{pq} \to G_q(V)$ are regular, surjective, holomorphic maps. Similarly

$$F = \{(x,y) \in \mathbb{P}(V) \times \mathbb{P}(V^*) \mid x \in \ddot{E}[y]\}$$

is a connected, smooth, compact, complex submanifold of $\mathbb{P}(V) \times \mathbb{P}(V^*)$ with dim F = 2n-1. The projections $\tau\colon F \to \mathbb{P}(V)$ and $\pi\colon F \to \mathbb{P}(V^*)$ are surjective, regular, (n-1)-fibering, holomorphic maps. If $a \in \mathbb{P}(V^*)$, the restriction

$$\tau\colon \pi^{-1}(a) \to \ddot{E}[a]$$

is biholomorphic.

The elements $x \in \mathbb{P}(\underset{p}{\Lambda} V)$ and $y \in \mathbb{P}(\underset{q}{\Lambda} V)$ are said to <u>intersect</u> if $\mathfrak{x} \in \mathbb{P}^{-1}(x)$ and $\mathfrak{y} \in \mathbb{P}^{-1}(y)$ exist such that $\mathfrak{x} \wedge \mathfrak{y} = 0$. If so, then $\mathfrak{x} \wedge \mathfrak{y} = 0$ for any possible choice of $\mathfrak{x}$ and $\mathfrak{y}$. If x and y do not intersect, then

$$x \wedge y = \mathbb{P}(\mathfrak{x} \wedge \mathfrak{y}) \in \mathbb{P}(\underset{p+q}{\Lambda} V)$$

is well defined where $\mathfrak{x} \in \mathbb{P}^{-1}(x)$ and $\mathfrak{y} \in \mathbb{P}^{-1}(y)$. Here $x \in G_p(V)$ and $y \in G_q(V)$ imply $x \wedge y \in G_{p+q-1}(V)$.

Let p and q be integers with $0 \leqq q \leqq p \leqq n$. Then $x \in \mathbb{P}(\underset{p}{\Lambda} V)$ and $y \in \mathbb{P}(\underset{q}{\Lambda} V^*)$ are said to <u>intersect</u> if $\mathfrak{x} \in \mathbb{P}^{-1}(x)$ and $\eta \in \mathbb{P}^{-1}(y)$ exist such that $\mathfrak{x} \llcorner \eta = 0$. If so, then $\mathfrak{x} \llcorner \eta = 0$ for any possible choice of $\mathfrak{x}$ and η. If x and y do not intersect, then

$$x \llcorner y = \mathbb{P}(\mathfrak{x} \llcorner \eta) \in \mathbb{P}(\underset{p-q}{\Lambda} V)$$

is defined. Observe $x \in \mathbb{P}(V)$ and $y \in \mathbb{P}(V^*)$ intersect if and only if $x \in \ddot{E}[a]$.

Now, assume that V is a hermitian vector space. If $x \in \mathbb{P}(\underset{p}{\Lambda}V)$ and $y \in \mathbb{P}(\underset{q}{\Lambda}V)$, take $\mathfrak{x} \in \mathbb{P}^{-1}(x)$ and $\mathfrak{y} \in \mathbb{P}^{-1}(y)$ and define

$$\| x : y \| = \| \mathfrak{x} : \mathfrak{y} \|$$

This definition does not depend on the choice of $\mathfrak{x}$ and $\mathfrak{y}$. Obviously, $\| x:y \| = 0$ if and only if x and y intersect.

If $p \geqq q$, if $x \in \mathbb{P}(\underset{p}{\Lambda}V)$ and $y \in \mathbb{P}(\underset{q}{\Lambda}V^*)$, take $\mathfrak{x} \in \mathbb{P}^{-1}(x)$ and $\eta \in \mathbb{P}^{-1}(y)$ and define

$$\| x;y \| = \| \mathfrak{x} ; \eta \|$$

The definition does not depend on the choice of $\mathfrak{x}$ and $\mathfrak{y}$.

The exterior derivative d splits into $d = \partial + \bar{\partial}$ and twists to

$$d^{\perp} = i(\partial - \bar{\partial}) = -d^c$$

On $\underset{p+1}{\Lambda} V$, define the norm $\tau = \tau_p$ by $\tau_p(\mathfrak{x}) = |\mathfrak{x}|^2$. Define

$$\upsilon_p = \frac{1}{4\pi} dd^c \tau_p = \frac{i}{2\pi} \partial\bar{\partial} \tau_p > 0 \qquad \text{on } V$$

$$\omega_p = \frac{1}{4\pi} dd^c \log \tau_p = \frac{i}{2\pi} \partial\bar{\partial}\log \tau_p \geqq 0 \quad \text{on } V - \{0\}.$$

Here υ_p is the euclidean volume element on each complex submanifold of $\Lambda_{p+1} V$ and ω_p is called the projective volume element.

The hermitian product on V defines a hermitian product on $\Lambda_{p+1} V$ which induces the Fubini-Study Kaehler metric on $\mathbb{P}(\Lambda_{p+1} V)$ whose fundamental form is denoted by $\ddot{\omega}_p$. Then

$$\omega_p = \mathbb{P}^*(\ddot{\omega}_p).$$

If $a \in G_p(V)$, then

$$\int_{\ddot{E}(a)} \ddot{\omega}_0^r = 1$$

Also

$$\int_{\mathbb{P}(V)} \ddot{\omega}_0^n = 1$$

Also denote $\upsilon = \upsilon_0$, and $\omega = \omega_0$ and $\ddot{\omega} = \ddot{\omega}_0$.

III. STOKES THEOREM ON COMPLEX SPACES

The theory will be developed on complex manifolds. However, it is convenient to have Stokes Theorem available on complex spaces. For compact support, it was proved by Lelong [24], for domains, it was proven by Tung [47].

Several preparations are needed.

a) Some notations on complex spaces.

Let X be a complex space. The set I(X) of irregular (= singular) points is a thin analytic subset of X. The set $\mathfrak{R}(X)$ of regular (= simple) points of X is open and dense in X. A patch α is a biholomorphic map $\alpha\colon U_\alpha \to U'_\alpha$ of an open subset $U_\alpha \neq \emptyset$ of X onto an analytic subset U'_α of an open subset G_α of $\mathbb{C}^{n_\alpha}$. Let $j_\alpha\colon U'_\alpha \to G_\alpha$ be the inclusion. Define

$$\alpha_0 = j_\alpha \circ \alpha : \mathfrak{R}(U_\alpha) \to G_\alpha \tag{3.1}$$

A patch α is said to be smooth if U'_α is open in $\mathbb{C}^{n_\alpha}$. Let $\mathfrak{P}_X$ be the set of all patches of X and let $\mathfrak{S}_X$ be the set of all smooth patches. If X is a manifold, a smooth patch will be usually called a patch.

b) Differential forms.

Let X be a complex manifold. Let $A_k(X)$ be the exterior graded algebra of all differentiable forms of class C^k on X. Let $A_k^m(X)$ be the set of forms of degree m, and let $A_k^{p,q}(X)$ be the set of forms of bidegree (p,q) in $A_k(X)$. Here $k = 0$ means continuous, $k \in \mathbb{N}$ or $k = \infty$ means k-times continuously differentiable, $k = \rho$ means realanalytic and $k = \omega$ means holomorphic.

Now, let X be a complex space. Then $A_k(\mathfrak{R}(X))$ is defined. The form $\chi \in A_k(\mathfrak{R}(X))$ is said to be of class C^k on X if and only if for every $a \in X$ a patch $\alpha \in \mathfrak{P}_X$ with $a \in U_\alpha$ such that there exists $\tilde{\chi} \in A_k(G_\alpha)$ with $\alpha_0^*(\tilde{\chi}) = \chi | \mathfrak{R}(U_\alpha)$. The set $A_k(X)$ of all forms of class C^k on X is an exterior subalgebra of $A_k(\mathfrak{R}(X))$. If χ and ψ are forms of class C^k on X, so are $\chi + \psi$ and $\chi \wedge \psi$ and $\overline{\chi}$. If $k \geqq 1$, then $d\chi$, and $\partial\chi$ and $\overline{\partial}\chi$ are forms of class C^{k-1}. Define

$$A_k^m(X) = A_k(X) \cap A_k^m(\mathfrak{R}(X))$$

$$A_k^{p,q}(X) = A_k(X) \cap A_k^{p,q}(\mathfrak{R}(X))$$

Let X and Y be complex spaces. Let $f: X \to Y$ be a holomorphic map. Take $\chi \in A_k(Y)$. Then one and only one $f^*(\chi) \in A_k(X)$ exists satisfying the following property.

(P) Let $\alpha \in \mathfrak{P}_X$ and $\beta \in \mathfrak{P}_Y$. Assume a holomorphic map $\tilde{f}: G_\alpha \to G_\beta$ exists with $\tilde{f} \circ j_\alpha \circ \alpha = j_\beta \circ \beta \circ f$. Assume $\tilde{\chi} \in A_k(G_\beta)$ exists with $\beta_0^*(\tilde{\chi}) = \chi | \mathfrak{R}(U_\beta)$. Then

$$f^*(\chi) | \mathfrak{R}(U_\alpha) = (\tilde{f} \circ j_\alpha \circ \alpha)^*(\tilde{\chi}).$$

The existence of f^* is not trivial, since $f(X) \subseteq I(Y)$ is possible. If χ and ψ are forms of class C^k on Y, then

$$f^*(\chi + \psi) = f^*(\chi) + f^*(\psi)$$

$$f^*(\chi \wedge \psi) = f^*(\chi) \wedge f^*(\psi)$$

$$f^*(\overline{\chi}) = \overline{f^*(\chi)}$$

If $k \geqq 1$, then

$$d\, f^*(\chi) = f^*(d\chi)$$

$$\partial\, f^*(\chi) = f^*(\partial\chi) \qquad \bar{\partial} f^*(\chi) = f^*(\bar{\partial}\chi)$$

If X, Y and Z are complex spaces, if $f: X \to Y$ and $g: Y \to Z$ are holomorphic and if $\chi \in A_k(Z)$, then

$$(g \circ f)^*(\chi) = f^*(g^*(\chi)).$$

A form $\chi \in A_k^{p,p}(X)$ is said to be non-negative, $\chi \geqq 0$, if and only if $\varphi^*(\chi) \geqq 0$ for every holomorphic map $\varphi: G \to X$ of an open subset $G \neq \emptyset$ of $\mathbb{C}^p$. If χ and ψ in $A_k^{p,p}(X)$ are non-negative, then $\chi + \psi \geqq 0$. If $\chi \in A_k^{p,p}(X)$ and $\psi \in A_k^{1,1}(X)$ are non-negative, then $\chi \wedge \psi \geqq 0$. The form $\chi \in A_k^{p,p}(X)$ is said to be strictly non-negative if $\chi \geqq 0$ and if $\chi \wedge \psi \geqq 0$ for all $0 \leqq \psi \in A_0^{q,q}(X)$.

c) Boundary manifolds.

Let M be an oriented, m-dimensional, differentiable manifold of class C^∞. Let G be an open subset of M. Let $\partial G = \overline{G} - G$ be the boundary of G. An orientation preserving diffeomorphism

$$\alpha: U_\alpha \to U'_\alpha \times U''_\alpha$$

of class C^∞ is said to be a boundary patch of G if and only if the sets U_α, U'_α, U''_α are open in M, $\mathbb{R}$ and $\mathbb{R}^{m-1}$ respectively.

Moreover, let $\pi: U'_\alpha \times U''_\alpha \to U'_\alpha$ be the projection. Then

$$\pi \circ \alpha < 0 \quad \text{on } U \cap G \qquad \pi \circ \alpha > 0 \quad \text{on } U - G$$

$$U \cap \partial G = (\pi \circ \alpha)^{-1}(0)$$

Let $\psi: U'_\alpha \times U''_\alpha \to U''_\alpha$ be the projection. The set dG of all point $x \in \partial G$ such that a boundary patch α with $x \in U_\alpha$ exists is either empty or an oriented, smooth, pure (m-1)-dimensional submanifold of class C^∞ such that for each boundary patch α of G, the map $\psi \circ \alpha: U_\alpha \cap \partial G \to U''_\alpha$ is an orientation preserving diffeomorphism of class C^∞. Obviously, $\partial G - dG$ is closed. The manifold dG is called the <u>boundary manifold</u> of G.

Let G be an open subset of the pure dimensional complex space X. The boundary manifold of G is defined by

$$dG = D(G \cap \mathfrak{R}(X)) = d\,\mathfrak{R}(G)$$

in the complex manifold $\mathfrak{R}(X)$.

<u>d) Hausdorff measure</u>.

Take $0 \leqq p \in \mathbb{R}$. Let $\mathfrak{z}^p$ be the p-dimensional Hausdorff-measure on $\mathbb{R}^m$. (Federer [12] p. 171). For $S \subseteq X$ define

$$\zeta(S) = \frac{1}{\Gamma(\frac{p}{2}+1)} \left(\tfrac{1}{2}\,\Gamma(\tfrac{1}{2})\right)^p (\operatorname{diam} S)^p$$

For each $\rho > 0$, let $\mathfrak{A}_\rho(S)$ be the set of all sequences

$K = \{K_\nu\}_{\nu\in\mathbb{N}}$ of subsets $K_\nu \subseteq X$ with diam $K_\nu \leqq \rho$ such that $S \subseteq \bigcup_{\nu\in\mathbb{N}} K_\nu$. Let $\Phi_\rho(S)$ be the infimum of all sums $\sum_{\nu\in\mathbb{N}} \zeta(K_\nu)$ with $\{K_\nu\}_{\nu\in\mathbb{N}} \in \mathfrak{K}_\rho(S)$. Then $\Phi_\rho(S)$ is a decreasing function of ρ. Define

$$\mathfrak{H}^p(S) = \lim_{\rho\to 0} \Phi_\rho(S)$$

as the p-dimensional Hausdorff measure of S. The measure $\mathfrak{H}^0$ counts the elements of S. The measure $\mathfrak{H}^m$ is the Lebesgue measure on $\mathbb{R}^m$.

A subset S of a complex space is said to have locally finite $\mathfrak{H}^p$-measure, if and only if for every patch $\alpha \in \mathfrak{P}_X$ and every compact subset K of U_α, the measure $\mathfrak{H}^p(\alpha(K \cap S)) < +\infty$.

The subset S of X is said to have zero $\mathfrak{H}^p$-measure if and only if for every patch $\alpha \in \mathfrak{P}_X$ the measure $\mathfrak{H}^p(\alpha(U_\alpha \cap S)) = 0$.

An open subset G of the pure m-dimensional complex space X is said to be Stokes domain if and only if $\overline{G}$ is compact, if ∂G has locally finite $\mathfrak{H}^{2m-1}$-measure and if $\partial G - dG$ has zero $\mathfrak{H}^{2m-1}$-measure.

e) Support on subsets.

Let $S \neq \emptyset$ be a closed subset of the complex space X. Let $\chi \in A_k(X)$. Then χ is said to vanish on S in a neighborhood of $a \in S$ if and only if a patch $\alpha \in \mathfrak{P}_X$ with $a \in U_\alpha$ and a form $\tilde{\chi} \in A_k(G_\alpha)$ exists such that $\alpha_0^*(\tilde{\chi}) = \chi | \mathfrak{R}(U_\alpha)$ and such that $\tilde{\chi} | \alpha(S \cap U_\alpha) = 0$. Let S_0 be the set of all

$a \in S$ such that χ vanishes on S in a neighborhood of a. The support of χ along S is defined by

$$\operatorname{supp}_S \chi = S - S_0$$

The usual support of χ on X is defined as

$$\operatorname{supp} \chi = \operatorname{supp}_X \chi$$

f) Stokes Theorem.

Let $G \neq \emptyset$ be an open subset of a pure m-dimensional complex space. Let E be a thin analytic subset of X with $E \supseteq I(X)$. Let $\chi \in A_1^{2m-1}(X)$ with $\overline{\chi} = \chi$. Assume that $K = \overline{G} \cap \operatorname{supp} \chi$ is compact. Define $T = \operatorname{supp}_{\partial G} \chi$. Suppose that $(T-E) - dG$ has zero $\mathfrak{H}^{2m-1}$-measure on $\mathfrak{R}(X)$. Assume that $K \cap (\partial G - E)$ has locally finite $\mathfrak{H}^{2m-1}$-measure on $\mathfrak{R}(X)$. Let $j\colon dG \to \mathfrak{R}(X)$ be the inclusion

$$\Gamma_+ = \{x \in dG \mid j^*(\chi) > 0\}$$

$$\Gamma_- = \{x \in dG \mid j^*(\chi) < 0\}.$$

If $\overline{\Gamma}_+ \cap \overline{\Gamma}_- = \emptyset$, then $j^*(\chi)$ is integrable over dG. If $j^*(\chi)$ is integrable over dG, then

$$\int_G d\chi = \int_{dG} \chi$$

If ∂G is compact and if ∂G has locally finite $\mathfrak{H}^{2m-1}$-measure, then $j^*(\chi)$ is integrable over dG. From this general Stokes theorem the following corollary is obtained.

Corollary to the Stokes Theorem.

Let $G \neq \emptyset$ be a Stokes domain on a pure m-dimensional complex space. Take $\chi \in A_1^{2m-1}(X)$. Then

$$\int_G d\chi = \int_{dG} \chi$$

Concerning the sign of the boundary differential, the following Lemma was proved in [36] Satz 4.5.

Lemma 3.1. Let $G \neq \emptyset$ be an open subset of the pure m-dimensional manifold space X. Let $j: dG \to X$ be the inclusion map. Take $a \in dG$. Let $\psi: X \to \mathbb{R}$ be a function of class C^1 on X. Assume that on open neighborhood U of a in X exists such that $\psi(x) \geqq \psi(a)$ for all $x \in U \cap \overline{G}$. Let $\chi \geqq 0$ be a continuous form of bidegree (m-1, m-1) on X. Then

$$j^*(d^\perp\psi \wedge \chi)(a) \geqq 0$$

For proofs for the statements of this section, see Tung [47]. See also Cowen [10], King [22] and Bloom-Herrera [3].

IV. MEROMORPHIC MAPS[1)]

Let X and Y be complex spaces. Let S be a thin analytic subset of X. Consider a holomorphic map

1) For general references see [36] and [2].

$$f: X - S \to Y$$

Let F be the closure in $X \times Y$ of the graph

$$F_0 = \{(x, f(x)) \mid x \in X - S\}.$$

Then F is called the <u>closed graph</u> of f. Let $\hat{f}$: $F \to Y$ and $\tilde{f}$: $F \to X$ be the projections. The map f is said to be <u>meromorphic</u> on X if and only if F is an analytic subset of $X \times Y$ and if $\tilde{f}$ is proper. (Remmert [28], Stein [33], and [2]). Adopt Stein's notation

$$f: X \rightharpoonup Y$$

for a meromorphic map. If $A \subseteq X$, the image is defined by

$$f(A) = \hat{f}(\tilde{f}^{-1}(A))$$

If $B \subseteq Y$, the counter image is defined by

$$f^{-1}(B) = \tilde{f}(\hat{f}^{-1}(B))$$

If B is analytic, then $f^{-1}(B)$ is analytic.

A largest open subset A of X exists such that f extends to a holomorphic map f_1: $A \to Y$. Then $\tilde{A} = \tilde{f}^{-1}(A)$ is open in F and $\tilde{f}$: $\tilde{A} \to A$ is biholomorphic. The set $I_f = X - A$ is analytic and nowhere dense. Obviously $I_f \subseteq S$ and $\tilde{A} \supseteq F_0$. The set I_f is called the <u>indeterminacy</u> of f. If X is a normal complex space, then $2 + \dim_x I_f \leqq \dim_x X$ for each $x \in I_f$.

Let M be a complex space. Let $\varphi: M \to X$ be a holomorphic map such that $S_0 = \varphi^{-1}(I_f)$ thin analytic subset of M. Then $f \circ \varphi: M - S_0 \to Y$ is a holomorphic map which is meromorphic on M. Also write $f \circ \varphi = \varphi^*(f)$.

Here meromorphic maps into projective spaces shall be considered. According to [2] Proposition 3.12, they can be represented locally by holomorphic vector functions. Let M be a pure m-dimensional complex space. Let V be a complex vector space of dimension n+1 with $n \geqq 0$. Let S be a thin analytic subset of M. Let

$$f: M - S \to \mathbb{P}(V)$$

be a holomorphic map. A holomorphic vector function $\mathfrak{w}: U \to V$ on an open subset U of M is said to be <u>representation of f</u> on U if and only if $\mathfrak{w}^{-1}(0)$ is thin analytic in U and

$$f(x) = \mathbb{P}(\mathfrak{w}(x))$$

for each $x \in U - S$ with $\mathfrak{w}(x) \neq 0$. The map f is meromorphic if and only if for every $a \in M$ an open neighborhood U of a and a representation $\mathfrak{w}: U \to V$ of f on U exists. A representation $\mathfrak{w}: U \to V$ is said to be simple if and only if $\mathfrak{w}(x) \neq 0$ for all $x \in U$. If $\mathfrak{w}: U \to V$ is simple, then $U \subseteq M - I_f$. If $a \in M - I_f$, a simple representation

$\mathfrak{w}: U \to V$ with $a \in U$ exists. If $\mathfrak{w}_\lambda: U_\lambda \to V$ are simple representations with $U_1 \cap U_2$, then a holomorphic function

$$h: U_1 \cap U_2 \to \mathbb{C} - \{0\}$$

exists such that $\mathfrak{w}_1 = h\,\mathfrak{w}_2$ on $U_1 \cap U_2$. The representation $\mathfrak{w}: U \to V$ is said to be *reducible* at $a \in U$, if and only if a holomorphic function h on an open neighborhood W of a in U and a holomorphic vector function $\mathfrak{v}: W \to V$ exist such that $\mathfrak{w} = h\,\mathfrak{v}$ on W and such that $h(a) = 0$. The representation $\mathfrak{w}$ is said to be *irreducible* if and only if $\mathfrak{w}$ is not reducible at each $a \in U$. Each simple representation is irreducible. If $a \in \mathfrak{R}(M)$, an irreducible representation $\mathfrak{w}: U \to V$ with $a \in U$ exists. If $a \in I(M)$ such an irreducible representation on a neighborhood may not exist. *This is one of the reasons why the theory will be developed on manifolds only*.

Now assume that M is a connected complex manifold of dimension m. Again, let $f: M \longrightarrow \mathbb{P}(V)$ be a meromorphic map. A representation $\mathfrak{w}: U \to V$ of V on an open subset U is irreducible, if and only if $\dim \mathfrak{w}^{-1}(0) \leqq m-2$. If $\mathfrak{w}: U \to V$ is an irreducible representation and if $\mathfrak{v}: W \to V$ is a representation with $W \cap U \neq \emptyset$, one and only one holomorphic function $h: W \cap U \to \mathbb{C}$ exists such that $\mathfrak{v} = h\,\mathfrak{w}$ on $U \cap W$. Moreover $\mathfrak{v}$ is irreducible on $W \cap U$ if and only if $h(x) \neq 0$ for all $x \in U \cap W$.

Let $\mathcal{O}_a$ be the ring of germs of holomorphic functions at a. Let $\mathfrak{m}_a$ be the maximal ideal. Let h be a holomorphic function in an open neighborhood of a. Then h defines germ h_a in $\mathcal{O}_a$. Assume $h_a \neq 0$. One and only one integer $p \geqq 0$ exists such that $h_a \in \mathfrak{m}^p$ but $h_0 \notin \mathfrak{m}^{p+1}$. The number $\delta_h(a)$ is called the zero multiplicity of h at a. The function δ_h is called the (zero) divisor of h.

Since M is a complex manifold, a divisor can be defined as a function $\nu: M \to \mathbb{Z}$ such that for every $a \in M$ an open, connected neighborhood U of a and holomorphic functions $g \not\equiv 0$ and $h \not\equiv 0$ on U exist such that $\nu | U = \delta_g - \delta_h$. The set of divisors on M is a module. A divisor ν is non-negative if and only if for every $a \in M$ an open neighborhood U of a and a holomorphic function $g \not\equiv 0$ exist such that $\nu | U = \delta_g$. The support $\gamma(\nu)$ of a divisor ν is an analytic subset of M, either empty or of pure dimension m-1. The function ν is locally constant on $\mathfrak{R}(\gamma(\nu))$.

Again, consider the meromorphic map $f: M \to \mathbb{P}(V)$. Let $\mathfrak{v}: U \to V$ be a representation. Take any $x_0 \in U$. An open connected neighborhood W of x_0 in U, an irreducible representation $\mathfrak{w}: W \to V$ and a holomorphic function $h \not\equiv 0$ on W exist such that $\mathfrak{v} | W = h \cdot \mathfrak{w}$. The number $\delta_{\mathfrak{v}}(x_0) = \delta_h(x_0)$ does not depend on the choice of h, $\mathfrak{w}$ and W. The function $\delta_{\mathfrak{v}}: U \to \mathbb{Z}$ is a non-negative divisor, called the greatest common divisor of $\mathfrak{v}$ (g.c.d.). Obviously, $\delta_{\mathfrak{v}} \equiv 0$ if and only if $\mathfrak{v}$ is irreducible.

Take $a \in \mathbb{P}(V^*)$. Then $f^{-1}(\ddot{E}[a]) = \gamma_f(a)$ is analytic. Either $\gamma_f(a) = \emptyset$ or $\gamma_f(a)$ pure (m-1)-dimensional or $\gamma_f(a) = M$ because M is connected. In the notation of [42], f is said to be adapted to a if $f^{-1}(\ddot{E}[a]) \neq M$. If f is adapted to every $a \in \mathbb{P}(V)$, then f is said to be generic, which is the case if and only if $f(M) \not\subseteq \ddot{E}[a]$ for each $a \in \mathbb{P}(V)$.

Take $a \in \mathbb{P}(V^*)$. Assume that f is adapted to a. Take $x \in M$. Let $\mathfrak{v}: U \to V$ be an irreducible, representation with $x \in U$. Take $\alpha \in \mathbb{P}^{-1}(a)$. Then $\alpha \circ \mathfrak{v}: U \to \mathbb{C}$ is a holomorphic function with $(\alpha \circ \mathfrak{v})^{-1}(0) = \gamma_f(a) \cap U$. The <u>intersection number</u> $\delta_f^a(x) = \delta_{\alpha \circ \mathfrak{v}}(x)$ of f with a at x does not depend on the choice of U, $\mathfrak{v}$, and α. The function $\delta_f^a: M \to \mathbb{Z}$ is a non-negative divisor called the <u>intersection divisor of f with a</u>. The support of δ_f^a is denoted by $\gamma_f(a)$.

<u>Lemma 4.1</u>. Let M be a connected, complex manifold of dimension m. Let V be a complex vector space of dimension n+1 with $n > 0$. Let $f: M \rightarrow \mathbb{P}(V)$ be a meromorphic map. Then a linear subspace W of V* with $W \neq V^*$ exists such that f is adapted to $a \in \mathbb{P}(V^*)$ if and only if $a \notin \mathbb{P}(W)$.

<u>Proof</u>. Let $\mathfrak{v}: U \to V$ be an irreducible representation of f on some open connected subset of M. Take $x_0 \in U$ with $\mathfrak{v}(x_0) \neq 0$. Define

$$W = \{\alpha \in V^* \mid \alpha \circ \mathfrak{v} \equiv 0\}.$$

Then W is a linear subspace of V*. Obviously, $a \in \mathbb{P}(W)$ if and only if f is not adapted to a, because M is connected. Since $\mathfrak{v}(x_0) \neq 0$, a vector $\alpha \in V^*$ with $\alpha(\mathfrak{v}(x_0)) \neq 0$ exists. Hence $\alpha \notin W$ and $W \neq V^*$; q.e.d.

Meromorphic maps $f: M \rightarrow \mathbb{P}(V)$ have been described by representations. Another description is provided by the hyperplane section bundle. The set

$$S(V) = \{(x, \mathfrak{z}) \in \mathbb{P}(V) \times V \mid \mathfrak{z} \in E(x)\}$$

is a closed, smooth, connected complex submanifold of $\mathbb{P}(V) \times V$. The projections $\sigma: S(V) \to V$ and $\pi: S(V) \to \mathbb{P}(V)$ are holomorphic with

$$\pi^{-1}(x) = \{x\} \times E(x)$$

Here $\pi: S(V) \to \mathbb{P}(V)$ is a holomorphic line bundle called the <u>hyperplane section bundle</u>. It is a subbundle of the trivial bundle $V_{\mathbb{P}(V)} = \mathbb{P}(V) \times V$. The restriction

$$\sigma: S(V) - \mathbb{P}(V) \times \{0\} \to V - \{0\}$$

is biholomorphic. The map σ is a σ-process in the sense of H. Hopf.

Let $\mathfrak{w}: U \to V$ be a holomorphic vector function on the open subset U of $\mathbb{P}(V)$ such that $\mathbb{P}(\mathfrak{w}(x)) = x$ if $\mathfrak{w}(x) \neq 0$ and $x \in U$. Then $s(x) = (x, \mathfrak{w}(x))$ defines a holomorphic section $s: U \to S(V)$ of $S(V)$ over U. Each holomorphic section of $S(V)$ over U can so be obtained. Hence the map $\mathfrak{w}$ itself is called a holomorphic section of U over V.

For each $0 \neq \alpha \in V^*$ consider the hyperplane $H_\alpha = \alpha^{-1}(1)$. For each $x \in \mathbb{P}(V) - \ddot{E}[\alpha]$ one and only one point $\mathfrak{w}_\alpha(x) \in E(x) \cap H_\alpha$ exists. Take $\mathfrak{x} \in \mathbb{P}^{-1}(x)$, then

$$\mathfrak{w}_\alpha(x) = \frac{\mathfrak{x}}{\alpha(\mathfrak{x})}$$

The map

$$\mathfrak{w}_\alpha : \mathbb{P}(V) - \ddot{E}[\alpha] \to V - \{0\}$$

is a holomorphic section of $S(V)$ over $\mathbb{P}(V) - \ddot{E}[\alpha]$. Since $\mathfrak{w}_\alpha(x) \neq 0$ for all $x \in \mathbb{P}(V) - \ddot{E}[\alpha]$, the section $\mathfrak{w}_\alpha$ is a holomorphic frame of $S(V)$ over $\mathbb{P}(V) - \ddot{E}[\alpha]$, providing a base in each fiber.

If $0 \neq \beta \in V^*$, a holomorphic function

$$g_{\alpha\beta} : \mathbb{P}(V) - (\ddot{E}(\alpha) \cup \ddot{E}(\beta)) \to \mathbb{C} - \{0\}$$

is defined by

$$g_{\alpha\beta}(x) = \frac{\beta(\mathfrak{z})}{\alpha(\mathfrak{z})} \quad \text{if} \quad \mathfrak{z} \in \mathbb{P}^{-1}(x)$$

Then

$$\mathfrak{w}_\alpha = g_{\alpha\beta}\, \mathfrak{w}_\beta \qquad \text{on } \mathbb{P}(V) - (\ddot{E}[\alpha] \cap \ddot{E}[\beta])$$

Hence $\{g_{\alpha\beta}\}_{(\alpha,\beta) \in \mathbb{P}(V^*) \times \mathbb{P}(V^*)}$ is a basic cocycle defining the hyperplane section bundle.

Let M be a complex space. Let $f: M \to \mathbb{P}(V)$ be a holomorphic map. Then

$$L = f^*(S(V)) = \{(x, \mathfrak{z}) \in M \times V \mid \mathfrak{z} \in E(f(x))\}$$

is an analytic subset of $M \times V$. Let $\pi: L \to M$ be the projection. Then $\pi: L \to M$ is a holomorphic line bundle and a subbundle of the trivial bundle $V_M = M \times V$. The line bundle L is the pullback of S(V) by f. The line bundle π: L $\rightarrow$ M is also called the hyperplane section bundle of f over M. A holomorphic section s: U $\rightarrow$ L can be identified with a holomorphic vector function $\mathfrak{v}$: U $\rightarrow$ V such that $\mathbb{P}(\mathfrak{v}(x)) = f(x)$ whenever $x \in U - \mathfrak{v}^{-1}(0)$. The identification is accomplished by

$$s(x) = (x, \mathfrak{v}(x)).$$

Let $\mathfrak{v}$: U $\rightarrow$ L be a holomorphic vector function such that $\mathfrak{v}^{-1}(0)$ is thin analytic in U. Then $\mathfrak{v}$ defines a holomorphic section of L over U if and only if $\mathfrak{v}$ is a representation of f over U. Moreover $\mathfrak{v}$ is a holomorphic frame of L over U if and only if $\mathfrak{v}$ is a simple representation. Hence another interpretation of the concept "representation of f" is obtained at least if f holomorphic.

The introduction of associated maps forces the consideration of meromorphic maps even if the original map is holomorphic. Therefore, the consideration of a meromorphic map

$$f: M \rightharpoonup \mathbb{P}(V)$$

cannot be avoided. The pullback f*(S(V)) is only defined over $M - I_f$. Because the Gauss-Bonnet formula for the line bundle will be applied to f*(S(V)), this deficiency is serious. As a remedy, M can be replaced by the graph F of f and f can be replaced by the holomorphic map $\hat{f}$: F $\rightarrow \mathbb{P}(V)$. This method works fine for the first main theorem. However, the deficit estimate uses associated maps f_p. Hence for each map f_p another resolution space F_p is needed. If M is a complex manifold, then dim $I_f \leqq m-2$. The line bundle $f^*(S(V))$ can be continued over M by Shiffman [32]. So the difficulty of the different resolution spaces can be avoided.

Without using Shiffman [32], the continuation can be constructed directly. Assume that M is a connected complex manifold of dimension m. Assume that f: $M \longrightarrow \mathbb{P}(V)$ is a meromorphic map. Let $\{\mathfrak{w}_\lambda: U_\lambda \to V\}_{\lambda \in \Lambda}$ be the family of all irreducible representations of f on M. Since $M = \mathfrak{R}(M)$, the family $\{U_\lambda\}_{\lambda \in \Lambda}$ is a covering of M. Define $L = f^*(S(V))$ as the pullback of f: $M - I_f \to \mathbb{P}(V)$ as a holomorphic line bundle over $M - I_f$. For each $\lambda \in L$, the vector function $\mathfrak{w}_\lambda$ defines a section L over $U_\lambda - I_\lambda$. Define

$$\Lambda[p] = \{(\lambda_0,\ldots,\lambda_p) \mid U_{\lambda_0} \cap \cdots \cap U_{\lambda_p} \neq \emptyset\}$$

If $(\lambda,\mu) \in \Lambda[1]$, a holomorphic function

$$g_{\lambda\mu}: U_\lambda \cap U_\mu \to \mathbb{C} - \{0\}$$

exists such that

$$\mathfrak{w}_\lambda = g_{\lambda\mu}\, \mathfrak{w}_\mu$$

on $U_\lambda \cap U_\mu$. Obviously, $g_{\lambda\mu} \cdot g_{\mu\lambda} = 1$. If $(\lambda,\mu,\rho) \in \Lambda[2]$, then

$$g_{\lambda\mu}\, g_{\mu\rho}\, g_{\rho\lambda} = 1$$

on $U_\lambda \cap U_\mu \cap U_\rho \neq \emptyset$. The cocycle $\{g_{\lambda\mu}\}_{(\lambda,\mu) \in \Lambda[1]}$ defines a holomorphic line bundle $\tilde{L}$ over M such that for each $\lambda \in \Lambda$ a holomorphic section $e_\lambda : U_\lambda \to \tilde{L}$ exists such that

$$e_\lambda = g_{\lambda\mu} e_\mu$$

on $U_\lambda \cap U_\mu$ whenever $(\lambda,\mu) \in \Lambda[1]$. A vector bundle isomorphism

$$\varphi : L \to \tilde{L} | (M - I_f)$$

is defined by the following procedure: Take $(x, \mathfrak{z}) \in L$. Take $\lambda \in \Lambda$ with $x \in U_\lambda$. Then $\mathfrak{z} = z_\lambda \mathfrak{w}_\lambda(x)$ because $\mathfrak{w}_\lambda(x) \neq 0$.

$$\varphi(x, \mathfrak{z}) = z_\lambda e_\lambda(x).$$

Obviously, $\varphi(x, \mathfrak{z})$ does not depend on the choice of λ. The bundle isomorphism φ identifies L with the restriction of $\tilde{L}$ to $M - I_f$ such that $\mathfrak{w}_\lambda$ is identified with e_λ. Under this identification, the family $\{\mathfrak{w}_\lambda\}_{\lambda \in \Lambda}$ of irreducible representations of f is precisely the family of holomorphic frames of $\tilde{L}$. A holomorphic vector function $\mathfrak{w} : U \to V$ identifies with a section of $\tilde{L}$ over U if and only if $\mathbb{P}(\mathfrak{w}(x)) = f(x)$ for each $x \in U$ with $\mathfrak{w}(x) \neq 0$; if $\mathfrak{w}^{-1}(0)$ is thin in U, then $\mathfrak{w}$ corresponds to a holomorphic section of $\tilde{L}$ over U if and only if $\mathfrak{w}$ is a representation of f over U. Therefore identify $L = \tilde{L} | (M - I_f)$ by φ and identify $e_\lambda = \mathfrak{w}_\lambda$ for all $\lambda \in \Lambda$. Observe, if $x \in I_f$, then $\tilde{L}_x \subseteq V$. Also $\mathfrak{w}_\lambda(x) = 0$ as a vector in V, but $\mathfrak{w}_\lambda(x) \neq 0$ as a value of the frame in $\tilde{L}_x$. Now, drop the ~ on L, such that $\tilde{L}$ itself is denoted by $L = f^*(S(V))$.

Let E be any holomorphic line bundle over M. Let $s: U \to E$ be a holomorphic section in E over the open subset $U \neq \emptyset$ of M. Assume that $s^{-1}(\boldsymbol{\sigma}) = \{x \mid s(x) = 0_x\}$ is thin analytic in U. Then s defines a non-negative divisor δ_s, <u>called the zero divisor of s</u>. Take any holomorphic frame $e: W \to E$ on an open subset W of U. Then $s = s_e \cdot e$ on W, where $s_e: W \to \mathbb{C}$ is a holomorphic function. Obviously, $s^{-1}(\boldsymbol{\sigma}) \mid W = s_e^{-1}(0)$. Define $\delta_s \mid W = \delta_{s_e}$. This definition does not depend on the choice of W and e.

If $\mathfrak{w}: U \to V$ is a representation of f over U, then $\mathfrak{w}$ is a holomorphic section in L. On $U \cap U_\lambda$ a holomorphic function h_λ exists such that $\mathfrak{w} = h_\lambda \mathfrak{w}_\lambda$. By definition $\delta_{\mathfrak{w}} \mid U \cap U_\lambda = \delta_{h_\lambda}$. Hence $\delta_{\mathfrak{w}}$ is the zero divisor of the section $\mathfrak{w}: U \to L$.

Let L* be the dual line bundle of L. It is given by the cocycle $\{g_{\nu\mu}\}_{(\mu,\nu) \in \Lambda[1]}$. Each element of L_x^* is a linear function on L_x. Take $\lambda \in \Lambda$. Then $\mathfrak{w}_\lambda: U_\lambda \to L$ is a holomorphic frame of L over U_λ. Let $\mathfrak{w}_\lambda^*: U_\lambda \to L^*$ be the dual frame. Then $\mathfrak{w}_\lambda^*(x)(\mathfrak{w}_\lambda(x)) = 1$ for all $x \in U_\lambda$. Write $\mathfrak{w}_\lambda^*(\mathfrak{w}_\lambda) = 1$. If $(\lambda,\mu) \in \Lambda[1]$, then

$$\mathfrak{w}_\lambda^* = \mathfrak{w}_\mu^* \, g_{\mu\lambda}$$

on $U_\lambda \cap U_\mu$.

Take $0 \neq \alpha \in V^*$. Then $\alpha \circ \mathfrak{w}_\lambda: U_\lambda \to \mathbb{C}$ is a holomorphic function. If $(\lambda,\mu) \in \Lambda[1]$, then

$$(\alpha \circ \mathfrak{w}_\lambda)\, \mathfrak{w}^*_\lambda = (\alpha \circ \mathfrak{w}_\mu)\, \mathfrak{w}^*_\mu$$

on $U_\lambda \cap U_\mu$. Hence one and only one holomorphic section

$$\alpha_f : M \to L^*$$

exists such that $\alpha_f | U_\lambda = (\alpha \circ \mathfrak{w}_\lambda)\, \mathfrak{w}^*_\lambda$. Obviously, $\delta_{\alpha_f} | U_\lambda = \delta_{\alpha} \circ \mathfrak{w}_\lambda = \delta^a_f$, if $a = \mathbb{P}(V^*)$. Hence

$$\delta_{\alpha_f} = d^a_f \quad \text{with } a = \mathbb{P}(\alpha).$$

Let $\mathfrak{w} : U \to V$ be a representation of f over the open subset $U \neq \emptyset$ of M. Take $\lambda \in \Lambda$ with $U_\lambda \cap U \neq \emptyset$. Then $\mathfrak{w} = h_\lambda \mathfrak{w}_\lambda$ on $U \cap U_\lambda$, where h_λ is a holomorphic function on $U_\lambda \cap U$. If $x \in U \cap U_\lambda$, then

$$(\alpha \circ \mathfrak{w})(x) = h_\lambda(x)\alpha(\mathfrak{w}_\lambda(x)) = h_\lambda(x)\ \alpha(\mathfrak{w}_\lambda(x))\, \mathfrak{w}^*_\lambda(x)(\mathfrak{w}_\lambda(x)) =$$

$$= \alpha_f(x)(\mathfrak{w}(x)) = \alpha_f(\mathfrak{w})(x)$$

Therefore

$$\alpha_f(\mathfrak{w}) = \alpha \circ \mathfrak{w}$$

$$\frac{|\alpha_f(\mathfrak{w})|}{|\alpha|\ |\mathfrak{w}|} = \| a;f \| \qquad \text{on } U - \mathfrak{w}^{-1}(0)$$

if $a = \mathbb{P}(\alpha)$.

Let E be a holomorphic vector bundle of fiber dimension q on M. A <u>hermitian metric along the fibers of E</u> is given by a function

$$h: E \oplus E \to \mathbb{C}$$

of class C^∞ such that $h_x: E_x \oplus E_x \to \mathbb{C}$ is a hermitian product on the vector space E_x. If $v \in E_x$ and $w \in E_x$, then $h(v,w) \in \mathbb{C}$ is defined. Write, $h(v) = h(v,v)$. Observe $h(v) > 0$ if $v \neq 0_x$. If $s: U \to E$ and $t: U \to E$ are sections, the functions $h(s,t)$ and $h(s)$ on U are defined by $h(s,t)(x) = h(s(x),t(x))$ and $h(s)(x) = h(s(x))$. If s and t are of class C^∞, then $h(s,t)$ and $h(s)$ are of class C^∞.

The hermitian metric h along the fibers of E defines the associated Chern forms $c_\mu(E,h)$ for $\mu = 0,1,\ldots,q$. Here $c_\mu(E,h)$ is a form of class C^∞ and bidegree (μ,μ) with $dc_\mu(E,h) = 0$. Also $c_0(E,h) = 1$. If $q = 1$, only the first Chern form $c_1(E,h)$ is essential and it can be computed easily: Let $e: U \to E$ be a holomorphic frame of E over U. Then $h(e) > 0$ is a function of class C^∞ on U. Then

$$c_1(E,h)|U = \frac{1}{4\pi}\, dd^c \log h(e).$$

Recall that $L|(M - I_f)$ is a subbundle of the trivial bundle $V_M = M \times V$ over $M - I_f$. The hermitian product on V defines a hermitian metric along the fibers of V_M which restricts to a hermitian metric ℓ along the fibers of $L|(M - I_f)$. If $\mathfrak{v}: U \to V$ and $\mathfrak{w}: U \to V$ are representations of f over the open subset $U \neq \emptyset$ of M, then $\mathfrak{v}$ and $\mathfrak{w}$ can be considered as sections of L over U. Then

$$\ell(\mathfrak{v}, \mathfrak{w}) = (\mathfrak{v} \mid \mathfrak{w})$$

$$\ell(\mathfrak{v}) = |\mathfrak{v}|^2$$

on $U - I_f$. Observe, that ℓ is not defined on the fibers L_x with $x \in I_f$.

Let h be a hermitian metric along the fibers of L. For $\lambda \in \Lambda$, the function $h_\lambda = h(\mathfrak{v}_\lambda)$ of class C^∞ on U_λ is defined. If $(\lambda, \mu) \in \Lambda[1]$, then

$$h_\lambda = |g_{\lambda\mu}|^2 \, h_\mu$$

$$|\mathfrak{v}_\lambda|^2 = |g_{\lambda\mu}|^2 \, |\mathfrak{v}_\mu|^2$$

on $U_\lambda \cap U_\mu$. Hence one and only one positive function $\| f \|_h$ of class C^∞ exists on $M - I_f$ such that

$$\| f \|_h^2 = \frac{h_\lambda}{|\mathfrak{v}_\lambda|^2} \qquad \text{on } U_\lambda - I_f$$

If $\mathfrak{v} : U \to V$ is a representation of f over the open subset $U \neq \emptyset$ of M, then

$$h(\mathfrak{v}) = \| f \|_h^2 \, |\mathfrak{v}|^2 \qquad \text{on } U - I_f$$

For $a \in \mathbb{P}(V^*)$, define

$$\| a;f \|_h = \frac{\| a;f \|}{\| f \|_h}$$

If $\mathfrak{w}: U \to V$ is a representation of f over U, and if $\alpha \in \mathbb{P}^{-1}(a)$, then

$$\| a;f \|_h^2 = \frac{|\alpha \circ \mathfrak{w}|^2}{|\alpha|^2 h(\mathfrak{w})}$$

on U . Hence $\| a;f \|_h^2$ is a function of class C^∞ on M. Observe

$$c_1(L,h) = \frac{1}{4\pi} dd^c \log h_\lambda$$

$$c_1(L, \ell) = \frac{1}{4\pi} dd^c \log |\mathfrak{w}_\lambda|^2 = f^*(\ddot{\omega})$$

$$c_1(L,h) - f^*(\ddot{\omega}_0) = \frac{1}{2\pi} dd^c \log \| f \|_h$$

Meromorphic functions can be considered as meromorphic maps into $\mathbb{P}(C^2) = \mathbb{C} \cup \{\infty\}$. Here $\mathbb{P}(\mathbb{C}^2)$ is identified with $\mathbb{C} \cup \{\infty\}$ by

$$\mathbb{P}(1,z) = z \qquad \text{if } z \in \mathbb{C}$$

$$\mathbb{P}(0,z) = 1 \qquad \text{if } 0 \neq z \in \mathbb{C}$$

A meromorphic function f on M is a meromorphic map $f: M \to \mathbb{P}(\mathbb{C}^2)$ such that $f^{-1}(\infty)$ is thin analytic. The set $\mathfrak{K}(M)$ of meromorphic functions on M is a field, since M is connected. If $f \in \mathfrak{K}(M)$ then $P_f = I_f \cup f^{-1}(\infty)$ is analytic and call the set of poles of f. Obviously $f: M - P_f \to \mathbb{C}$ is holomorphic.

The dual space $\mathbb{P}(\mathbb{C}^{2*})$ is identified with $\mathbb{C} \cup \{\infty\}$ by

$$\mathbb{P}(-z,1) = z \qquad \text{if } z \in \mathbb{C}$$

$$\mathbb{P}(z,0) = \infty \qquad \text{if } 0 \neq z \in \mathbb{C}$$

Take $a \in \mathbb{P}(\mathbb{C}^{2*})$. Take $\alpha = (\alpha_0,\alpha_1) \in \mathbb{P}^{-1}(a)$. If $\alpha_1 \neq 0$, then $a = -\dfrac{\alpha_0}{\alpha_1}$ under this identification. If $\alpha_1 = 0$, then $a = \infty$. Take $f \in \mathfrak{K}(M)$. Let $\mathfrak{w} = (g,h)\colon U \to \mathbb{C}^2$ be an irreducible representation of the meromorphic map $f\colon M \to \mathbb{P}(\mathbb{C}^2)$ on the open subset $U \neq \emptyset$ of M. Then $g \cdot f = h$ on U and $g^{-1}(0)$ is thin in U. Observe

$$\alpha \circ \mathfrak{w} = (\mathfrak{w};\alpha) = \alpha_0 g + \alpha_1 h$$

Assume f is adapted to a, i.e. $f \not\equiv a$ on M. Then

$$\delta_f^a = \delta_{\alpha_0 g + \alpha_1 h}$$

on U. If $\alpha_1 \neq 0$ and if f is holomorphic at $x \in U$, then $\delta_{\alpha_0 g + \alpha_1 h}(x) = \delta_{f-a}(x)$. If $x \in P_f \cap U - I_f$, then $g(x) \neq 0 = h(x)$ and $\delta_f^a(x) = 0$. If $x \in I_f$, then $g(x) = h(x) = 0$ and $\delta_f^a(x) > 0$. If $\alpha_1 = \infty$, then $a = \infty$ and $\delta_f^\infty(x) = \delta_g(x)$. Hence δ_f^a is <u>the a-divisor of the meromorphic function f</u> for $a \in \mathbb{C} \cup \{\infty\}$. If $f^{-1}(0)$ is thin, define

$$\delta_f = \delta_f^0 - \delta_f^\infty ,$$

as the divisor of f on M. Observe

$$\delta_{f\cdot g} = \delta_f + \delta_g$$

$$\delta_{f+g} \geqq \text{Min } (\delta_g, \delta_g)$$

The meromorphic function f on M is holomorphic if and only if $\delta_f \geqq 0$.

If $f \not\equiv 0$ is meromorphic on M, define

$$\gamma_f = \text{supp } \delta_f = \gamma_f(0) \cup \gamma_f(\infty)$$

V. THE FIRST MAIN THEOREM

The following general assumptions shall be made.

(A1) Let M be a connected complex manifold of dimension m.

(A2) Let $\chi \geqq 0$ be a non-negative, differential form of class C^∞ and of bidegree (m-1, m-1) on M such that $d\chi = 0$.

Lemma 3.1 together with [34] Satz 6.2 and Satz 6.3 imply

Theorem 5.1. Assume (A1) and (A2). Let G be a Stokes domain in M. Let ψ be a function of class C^∞ on M such that $\psi|\overline{G} \geqq 0$ and $\psi|\partial G = 0$. Let $f \not\equiv 0$ be a meromorphic function on M. Then

$$\int_{dG} \log|f| \; d^\perp\psi \wedge \chi = \int_G d \log|f| \wedge d^\perp\psi \wedge \chi + \int_G \log|f| dd^\perp\psi \wedge \chi$$

$$\int_G d\psi \wedge d^\perp \log|f| \wedge \chi = 2\pi \int_{G \cap \gamma_f} \delta_\rho \, \psi \, \chi$$

where all integrals exist.

Observe

$$d\psi \wedge d^\perp \log |f| \wedge \chi = d \log |f| \wedge d^\perp\psi \wedge \chi$$

If $j: dG \to M$ is the inclusion, $\psi \geqq 0$ on $\overline{G}$ and $\psi = 0$ on dG imply

$$j^*(d^\perp\psi \wedge \chi) \geqq 0$$

A triple $\mathfrak{G} = (G, g, \psi)$ is said to be a <u>bump</u> if and only if the following conditions are satisfied.

1. The set g is either empty or a Stokes domain on M.
2. The set $G \neq \emptyset$ is a Stokes domain on M with $G \supset \overline{g}$.
3. The continuous function ψ is non-negative on M with $\psi(x) = 0$ for all $x \in M - G$.
4. The restriction $\psi|(\overline{G} - g)$ is the restriction of some function $\tilde{\psi}$ of class C^∞ on M such that $\tilde{\psi} \geqq \psi$ on G.
5. If $g \neq 0$, then $\psi|\overline{g} = R \geqq 0$ is constant and $R \geqq \psi \geqq 0$ holds on M.

If $g = \emptyset$, the bump $\mathfrak{G}$ is called <u>simple</u>.

Let $\mathfrak{G} = (G, g, \psi)$ be a bump on M. Let $j_G: dG \to M$ and $j_g: dg \to M$ (if $g \neq \emptyset$) be the inclusion maps. Let $\tilde{\psi}$ be a function of class C^∞ on M. Then $j_G^* (d^\perp\tilde{\psi})$ and $j_g^* (d^\perp\tilde{\psi})$ do not depend on the choice of $\tilde{\psi}$ and are denoted by $j_G^*(d^\perp\psi)$ and $j_g^*(d^\perp\psi)$ respectively. Observe

$$j_G^*(d^\perp\psi \wedge \chi) \geqq 0 \qquad j_g^*(d^\perp\psi \wedge \chi) \geqq 0$$

by Lemma 3.1. Now the Jensen Formula is easily obtained.

Theorem 5.2. (Jensen Formula)

Assume (A1) and (A2). Let $\mathfrak{G} = (G,g,\psi)$ be a bump on M. Let $f \not\equiv 0$ be a meromorphic function on M. Then

$$\frac{1}{2\pi}\int_{dG}\log|f|\ d^\perp\psi \wedge \chi - \frac{1}{2\pi}\int_{dg}\log|f|\ d^\perp\psi \wedge \chi - \frac{1}{2\pi}\int_{G-\overline{g}}\log|f|dd^\perp\psi \wedge \chi$$

$$= \int_{\gamma_f \cap G} \delta_f\ \psi\ \chi$$

Proof. Let $\tilde{\psi}$ be a function of class C^∞ on M such that $\tilde{\psi}(x) = \psi(x)$ if $x \in \overline{G} - g$ and such that $\tilde{\psi} \geqq R$ on $\overline{g}$ if $\overline{g} \neq \emptyset$. The formulas of Theorem 5.1 imply

$$\int_{dG}\log|f|\ d^\perp\tilde{\psi} \wedge \chi = 2\pi \int_{G\cap\gamma_f} \delta_f\tilde{\psi}\chi + \int_G \log|f|\ dd^\perp\tilde{\psi} \wedge \chi$$

$$\int_{dg}\log|f|d^\perp(\tilde{\psi}-R) \wedge \chi = 2\pi \int_{g\cap\gamma_f} \delta_f(\tilde{\psi}-R)\chi$$

$$+ \int_g \log|f|\ dd^\perp(\tilde{\psi}-R) \wedge \chi$$

Observe $d^\perp R = 0$ and $\tilde{\psi}|\overline{G}-g = \psi|\overline{G}-g$ and $\psi|\overline{g} = R$. Substraction implies the Theorem;

q.e.d.

Theorem 5.3. The integrated Gauss-Bonnet formula.

Assume (A1) and (A2). Let E be a holomorphic line bundle on M. Let h be a hermitian metric along the fibers of E. Let $s\colon M \to E$ be a holomorphic section in E different from the zero section. Let $\mathfrak{G} = (G, g, \psi)$ be a bump on M. Then

$$\int_{\gamma(\delta_s)\cap G} \delta_s \psi \chi = \frac{1}{4\pi}\int_{dG} \log h(s)\, d^{\perp}\psi \wedge \chi - \frac{1}{4\pi}\int_{dg} \log h(s)\, d^{\perp}\psi \wedge \chi$$

$$- \frac{1}{4\pi}\int_{G-\overline{g}} \log h(s)\, dd^{\perp}\psi \wedge \chi$$

$$- \int_{G} \psi\, c_1(E,h) \wedge \chi$$

Proof. At first, assume that there is a holomorphic frame $v\colon U \to E$ of E on an open neighborhood U of G. Then $s = f \cdot v$ where $f\colon U \to \mathbb{C}$ is a holomorphic function on U with $\delta_f = \delta_s | U$. Also $h(v) = k > 0$ is a function of class C^∞ on U. Moreover

$$h(s) = k\, |f|^2$$

on U. Stokes' Theorem for the Stokes domain $G - \overline{g}$ implies

$$\int_{dG} \log k\, d^{\perp}\psi \wedge \chi - \int_{dg} \log k\, d^{\perp}\psi \wedge \chi - \int_{G-\overline{g}} \log k\, dd^{\perp}\psi \wedge \chi$$

$$= \int_{G-\overline{g}} d \log k \wedge d^{\perp}\psi \wedge \chi$$

$$= \int_{G-\overline{g}} d\psi \wedge d^{\perp} \log k \wedge \chi$$

$$= -R \int_{dg} d^{\perp} \log k \wedge \chi - \int_{G-g} \psi \, dd^{\perp} \log k \wedge \chi$$

$$= -R \int_{g} dd^{\perp} \log k \wedge \chi - \int_{G-g} \psi \, dd^{\perp} \log k \wedge \chi$$

$$= 4\pi \int_{G} \psi \, c_1(E,h) \wedge \chi$$

Divide this identity by 4π and add it to the Jensen formula for f. This yields the Gauss-Bonnet formular in this special case.

Now, assume that $\mathfrak{G}$ is a simple bump, i.e., $g = \emptyset$. Take $a \in \overline{G}$. An open neighborhood U(a) of a shall be constructed such that the Gauss-Bonnet formular holds for ψ replaced by $\lambda\psi$ where $\lambda \geqq 0$ is any function of class C^{∞} on M with compact support in U(a).

Let $\alpha \in \mathfrak{T}_M$ be a (smooth) patch on M with $a \in U_\alpha$ and $\alpha(a) = 0$. Then $r_0 > 0$ exists such that $\mathbb{C}^m(r_0) \subset U'_\alpha$ and such that a holomorphic frame v of E on $U_0 = \alpha^{-1}(\mathbb{C}^m(r_0)) \cap G$ exists. For each $r \in \mathbb{R}(0,r_0)$, define $G_r = \alpha^{-1}(\mathbb{C}^m(r)) \cap G$. Then $G_r \neq \emptyset$ is open and $\overline{G}_r$ is compact in U_0. Obviously, ∂G_r has locally finite $\mathfrak{H}^{2m-1}$ measure. A number $r \in \mathbb{R}(0,r_0)$ exists such that $\partial G_r - dG_r$ has zero $\mathfrak{H}^{2m-1}$-measure. Hence G_r is a Stokes domain. Define $U(a) = \alpha^{-1}(\mathbb{C}^m(r))$. Then U(a) is an open neighborhood of a. Let $\lambda \geqq 0$ be a function of class C^{∞} on M with compact support in U(a). Then $(G_r, \emptyset, \lambda\psi)$ is a bump. The first part of the proof implies

$$\gamma(\ _0) \int_{\cap G_r} s^{\lambda\psi\chi} = \frac{1}{4\pi} \int_{dG_r} \log h(s)\, d\,(\lambda\psi) \wedge \chi$$

$$- \frac{1}{4\pi} \int_{G_r} \log h(s) \, dd^{\perp} (\lambda\psi) \wedge \chi$$

$$- \int_{G_r} \lambda\psi \, c_1(E,h) \wedge \chi$$

Because λ has compact support in $U(a)$, it is possible to replace G_r by G in this formula.

A finite covering $U(a_1),\dots,U(a_p)$ of $\overline{G}$ exists. Take non-negative functions $\lambda_1,\dots,\lambda_p$ of class C^∞ on M such that λ_μ has compact support in $U(a_\mu)$ for $\mu = 1,\dots,p$ and such that $\lambda_1 + \dots + \lambda_p = 1$ in an open neighborhood of $\overline{G}$. Apply the last formula for λ_μ and add over $\mu = 1,\dots,p$. This gives the Gauss-Bonnet formula for simple bumps. The Gauss-Bonnet formula for bumps (G,g,ψ) with $g \neq \emptyset$ is easily obtained by the same procedure as in the proof of Theorem 5.2.

The following additional general assumptions are made.

(A3) Let V be a hermitian vector space of dimension $n+1$ with $n > 0$.

(A4) Let $f: M \longrightarrow \mathbb{P}(V)$ be a meromorphic map.

Theorem 5.5. (First Main Theorem Preliminary version). Assume (A.1)-(A4). Let $\mathfrak{L} = (G,g,\psi)$ be a bump on M. Take $a \in \mathbb{P}(V^*)$. Assume that f is adapted to a. Let h be a hermitian metric along the fibers of $L = f^*(S(V))$. Then the integrals

$$N_f(G,a) = \int_{G \cap \gamma_f(a)} \delta_f^a \, \psi \, \chi \geqq 0$$

$$m_f(dG,a,h) = \frac{1}{2\pi} \int_{dG} \log \frac{1}{||f;a||_h} \; d^{\perp}\psi \wedge \chi$$

$$m_f^0(dg,a,h) = \frac{1}{2\pi} \int_{dg} \log \frac{1}{||f;a||_h} \; d^{\perp}\psi \wedge \chi$$

$$D_f(G,a,h) = \frac{1}{2\pi} \int_{G-\overline{g}} \log \frac{1}{||f;a||_h} \; dd^{\perp}\psi \wedge \chi$$

$$T_f(G,h) = \int_G \psi \; c_1(E;h) \wedge \chi$$

exist and

$$N_f(G,a) + m_f(dG,a,h) - m_f(dg,a,h) - D_f(G,a,h) = T_f(G,h).$$

Remark. Although this identity gives a balancing statement, it is deficient by its dependency on h which does not permit to determine the signs of the integrals depending on h. These deficiencies will be removed soon.

Proof. Take $\alpha \in \mathbb{P}^{-1}(a)$ with $|\alpha| = 1$. Then $\alpha_f\colon M \to L^*$ is a holomorphic section. Let h^* be the dual hermitian metric to h. Then h^* is a hermitian metric along the fibers of L^*. Let $\mathfrak{v} : U \to V$ be an irreducible representation of f on the open subset $U \neq \emptyset$ of M. Then $\mathfrak{v}$ can be considered to be a holomorphic frame of L over U. Let $\mathfrak{v}^*$ be the dual frame. Then $h^*(\mathfrak{v}^*)\, h(\mathfrak{v}) = 1$ on U. Now

$$\alpha_f = (\alpha \circ \mathfrak{v}) \cdot \mathfrak{v}^*$$

on U. Therefore

$$h^*(\alpha_f) = \frac{|\alpha \circ \mathfrak{w}|^2}{h(\mathfrak{w})} = ||a;f||_h^2$$

on U. Therefore $h^*(\alpha_f) = ||a;f||_h^2$. Observe

$$c_1(L^*,h^*) = \frac{1}{4\pi} dd^c \log h^*(\mathfrak{w}^*) = -\frac{1}{4\pi} dd^c \log h(\mathfrak{w}) = -c_1(L,h)$$

on U. Hence $c_1(L^*,h^*) = -c_1(L,h)$. The application of Theorem 5.3 to L^*, h^* and α_f yields Theorem 5.6.

q.e.d.

Lemma 5.6. Assume (A1) and (A2). Let $\mathfrak{G} = (G,g,\psi)$ be a bump on M. Let h be a meromorphic function on an open subset U of M such that $h^{-1}(0)$ is thin in U. Let K be a compact subset of $U \cap \overline{G}$. Then the integrals

$$\int_{K \cap dG} \log |h| \, d^{\perp}\psi \wedge \chi$$

$$\int_{K \cap dg} \log |h| \, d^{\perp}\psi \wedge \chi$$

exist.

Proof. It suffices to prove the existence of the first integral. Take $a \in K \cap \partial G$. As in the proof Theorem 5.3 construct a patch $\alpha \in \mathfrak{T}_M$ with $a \in U(a) \subseteq U_\alpha \cap U$ such that $G_r = U(a) \cap G$ is a Stokes domain. Let W be an open neighborhood of a such that $\overline{W}$ is compact and contained in U(a). Take a non-negative function of class C^∞ on M such that $\lambda|\overline{W} = 1$ and such that λ has compact support in U(a). Then $(G_r, \emptyset, \lambda\psi)$ is a bump on M. According to Theorem 5.2 the integral

$$\int_{dG_r} \log |h| \; d^{\perp}(\lambda\psi) \wedge \chi$$

exists. Now, $W \cap dG \subseteq dG_r$ and $\lambda = 1$ on W. Hence

$$\int_{W \cap dG} \log |h| \; d^{\perp}\psi \wedge \chi$$

exists. Finitely many open neighborhoods W cover ∂G. Hence

$$\int_{dG} \log |h| \; d^{\perp}\psi \wedge \chi$$

exists.

Lemma 5.7. Assume (A1) - (A3). Let $\mathfrak{G} = (G, g, \psi)$ be a bump on M. Let $\mathfrak{w} : U \to V$ be a holomorphic vector function on the open subset U of M such that $\mathfrak{w}^{-1}(0)$ is thin in U. Let K be a compact subset of $U \cap \overline{G}$. Then the integrals

$$\int_{K \cap dG} \log |\mathfrak{w}| \; d^{\perp}\psi \wedge \chi \quad \text{and} \quad \int_{K \cap dg} \log |\mathfrak{w}| \; d^{\perp}\psi \wedge \chi$$

exist.

Proof. W.l.o.g. it can be assumed that U is connected. An orthonormal base $\mathfrak{n}_0, \ldots, \mathfrak{n}_n$ of V exist such that

$$\mathfrak{w} = \sum_{\mu=0}^{k} v_{\mu} \mathfrak{n}_{\mu}$$

where $0 < k \leqq n$ and where $v_{\mu} \not\equiv 0$ on U. Then

$$\log^{+} |\mathfrak{w}| = \frac{1}{2} \log^{+} \sum_{\mu=1}^{k} |v_{\mu}|^{2} \leqq \frac{1}{2} \sum_{\mu=1}^{k} \log^{+} |v_{\mu}|^{2} + \frac{1}{2} \log n$$

$$\leqq \sum_{\mu=1}^{k} |\log |v_\mu|| + \frac{1}{2} \log n$$

$$\log^+ \frac{1}{|\mathfrak{w}|} \leqq \log^+ \frac{1}{|v_0|} \leqq |\log |v_0||$$

By lemma 5.7

$$\int_{K \cap dg} |\log |v_\mu|| \, d^\perp\psi \wedge \chi$$

exists for $\mu = 0,\ldots,k$. Also $\int_{K \cap dG} \frac{1}{2} \log n \, d^\perp\psi \wedge \chi$ exists. Hence $\int_{K \cap dG} |\log |\mathfrak{w}|| \, d^\perp\psi \wedge \chi$ exists. q.e.d.

Define

$$\mathfrak{a}_f = \{a \in \mathbb{P}(V^*) \mid f \text{ is adapted to } a\}.$$

Theorem 5.8. First Main Theorem.

Assume (A1) - (A4). Let (G,g,ψ) be a bump on M. If $a \in \mathfrak{a}_f$, then

$$N_f(G,a) = \int_{G \cap \gamma_f(a)} \delta_f^a \, \psi\chi \geqq 0$$

$$m_f(dG,a) = \frac{1}{2\pi} \int_{dG} \log \frac{1}{||a;f||} \, d^\perp\psi \wedge \chi \geqq 0$$

$$m_f^0(dg,a) = \frac{1}{2\pi} \int_{dg} \log \frac{1}{||a;f||} \, d^\perp\psi \wedge \chi \geqq 0$$

$$D_f(G,a) = \frac{1}{2\pi} \int_{G-\overline{g}} \log \frac{1}{||a;f||} \, dd^\perp\psi \wedge \chi$$

exist. Moreover,

$$T_f(G) = N_f(G,a) + m_f(dG,a) - m_f^0(dg,a) - D_f(G,a)$$

is a constant function of a on $\mathfrak{A}_f$. If h is any hermitian metric along the fibers of the line bundle $L = f^*(S(V))$, then

$$T_f(G) = \frac{1}{2\pi}\int_\Gamma \log\frac{1}{||f||_h}\, d^\perp\psi \wedge \chi - \frac{1}{2\pi}\int_\gamma \log\frac{1}{||f||_h}\, d^\perp\psi \wedge \chi$$

$$- \frac{1}{2\pi}\int_G \log\frac{1}{||f||_h}\, d^\perp\psi \wedge \chi + \int_G \psi\, c_1(L,h) \wedge \chi$$

where all integrals exist.

Proof. Take $x_0 \in \partial G$. An irreducible representation $\mathfrak{w}: U \to V$ on an open neighborhood U of x_0 exists. Let W be an open neighborhood of x_0, such that $\overline{W}$ is compact and contained in V. Then

$$||f||_h^2 = \frac{h(\mathfrak{w})}{|\mathfrak{w}|^2}$$

on U. Here $h(\mathfrak{w}) > 0$ has class C^∞. By Lemma 3.7 the integral

$$\int_{W \cap dG} \log\frac{h(\mathfrak{w})}{|\mathfrak{w}|^2}\, d^\perp\psi \wedge \chi$$

exists. Finitely many such neighborhoods W cover ∂G. Hence

$$\int_{dG} \log ||f||_h\, d^\perp\psi \wedge \chi$$

exists. Similarly the existence of

$$\int_{dg} \log ||f||_h \, d^{\perp}\psi \wedge \chi$$

is obtained. Trivially $\int_{G-g} \log ||f||_h \, dd^{\perp}\psi \wedge \chi$ exists. Because $||a;f|| = ||a;f||_h \, ||f||_h$, Theorem 5.5 implies

$$N_f(G,a) + m_f(dG,a) - m_f^0(dg,a) - D_f(G,a)$$

$$= T_f(G,h) + \frac{1}{2\pi}\int_{dG} \log \frac{1}{||f||_h} d^{\perp}\psi \wedge \chi - \frac{1}{2\pi}\int_{dg} \log\frac{1}{||f||_h} d^{\perp}\psi \wedge \chi$$

$$- \frac{1}{2\pi}\int_{G-g} \log \frac{1}{||f||_h} dd^{\perp}\psi \wedge \chi = T_f(G)$$

for all $a \in \mathfrak{a}_f$. The first expression does not depend on h. The second expression does not depend on a. Hence $T_f(G)$ is independent of a and h;

q.e.d.

Hence $N_f(G,a)$ is the <u>valence function</u>, $m_f(dG,a)$ and $m_f^0(dg,a)$ are the <u>compensation functions</u> and $D_f(G,a)$ is the <u>deficit</u>. The invariant $T_f(G)$ is the <u>characteristic function</u>. Define also

$$n_f(G,a) = \int_{G \cap \gamma_f(a)} \delta_f^a \chi$$

as the <u>counting function</u> for $a \in \mathfrak{a}_f$.

<u>Theorem 5.9</u>. Assume (A1) - (A4). Let (G,g,ψ) be a bump on M. Let $\mathfrak{w} : U \to V$ be a representation of f on the open neighborhood U of $\overline{G}$. Then

$$T_f(G) = \frac{1}{2\pi}\int_{dG} \log|\mathfrak{w}|\; d^{\perp}\psi \wedge \chi - \frac{1}{2\pi}\int_{dg} \log|\mathfrak{w}|\; d^{\perp}\psi \wedge \chi$$

$$- \frac{1}{2\pi}\int_{G-g} \log|\mathfrak{w}|\; dd^{\perp}\psi \wedge \chi - \int_{\gamma(\delta_{\mathfrak{w}}) \cap G} \delta_{\mathfrak{w}}\, \psi\chi$$

Proof. W.l.o.G, it can be assumed that U is connected. Consider $\mathfrak{w}$ to be a section in $L = f^*(S(V))$. Theorem 5.3 implies

$$\int_{G \cap \gamma(\delta\quad)} \delta_{\mathfrak{w}}\, \psi\chi = \frac{1}{4\pi}\int_{dG} \log h(\mathfrak{w})\; d^{\perp}\psi \wedge \chi - \frac{1}{4\pi}\int_{dg} \log h(\mathfrak{w})\; d^{\perp}\psi \wedge \chi$$

$$- \frac{1}{4\pi}\int_{G-g} \log h(\mathfrak{w})\; dd^{\perp}\psi \wedge \chi - T_f(G,h)$$

Also

$$||f||_h^2 = \frac{h(\mathfrak{w})}{|\mathfrak{w}|^2}$$

Therefore Theorem 3.8 implies Theorem 3.9, q.e.d.

If $f\colon M \to \mathbb{P}(V)$ is a holomorphic map, then the metric on V_M defined by the hermitian product on M defines a hermitian metric ℓ along the fibers on M. Then $||f||_{\ell} = 1$ and $c_1(L, \ell) = f^*(\ddot{\omega}_0)$. Hence the First Main Theorem implies

$$T_f(G) = \int_G \psi\, f^*(\ddot{\omega}_0) \wedge \chi$$

This integral representation of T_f also holds if f is meromorphic. Two proofs shall be offered. The first uses approximation and Stokes theorem. The second uses the classical method of integral averages.

Lemma 5.10. Assume (A1). Let $f_1, \ldots, f_n$ be holomorphic functions on M which are coprime at every point of M. Assume $n \geqq 3$. Take $x_0 \in M$. Then complex numbers $a_{\mu\nu} \in \mathbb{C}$ for $\mu, \nu = 1, \ldots, n$ exist such that $\det(a_{\mu\nu}) \neq 0$ and such that the functions $h_\mu = a_{\mu 1} f_1 + \ldots + a_{\mu n} f_n$ are pairwise coprime at x_0 for $\mu = 1, \ldots, n$.

Proof. The results of [2] on the rank of a holomorphic map will be used. W.l.o.g. it can be assumed that M is an open, connected subset of $\mathbb{C}^m$. Let $G = GL(\mathbb{C}^n)$ be the set of all n to n matrices A with coefficients in $\mathbb{C}$ such that $\det A \neq 0$. Here G is considered as an open, connected, dense subset of $\mathbb{C}^{n^2}$. On $M \times G$ define a holomorphic function H_μ by

$$H_\mu(x, A) = \sum_{\nu=1}^{n} a_{\mu\nu} f_\nu(x)$$

where $x \in M$ and

$$A = \begin{pmatrix} a_{11}, \ldots, & a_{1n} \\ \vdots & \vdots \\ a_{n1}, \ldots, & a_{nn} \end{pmatrix} \in G$$

Define $H_{\mu\nu} = (H_\mu, H_\nu): M \times G \to \mathbb{C}^2$ as the junction of H_μ and H_ν.

At first, it will be shown, that the functions H_μ and H_ν are coprime at every point of $M \times G$ if $\mu \neq \nu$. For, assume they are not coprime at some point $(x_1, A_1) \in M \times G$. Then open, connected neighborhoods U_1 if x_1 in M and W_1 of A_1 in G and holomorphic functions g, u_μ, u_ν on $U_1 \times W_1$

exist such that g is prime at (x_1, A_1) such that

$$g^q u_\mu = H_\mu \qquad g^p u_\nu = H_\nu$$

on $U_1 \times W_1$ and such that g does not divide u_μ and also not u_ν at (x_1, A_1). Here $q \in \mathbb{N}$ and $p \in \mathbb{N}$ are positive integers. Take the partial derivative for $a_{\mu\rho}$. Then

$$g^{q-1}(u_{\mu a_{\mu\rho}} g + q\, g_{a_{\mu\rho}} u) = f_\rho$$

for $\rho = 1,\ldots,n$. If $q > 1$, then g divides each f_ρ at (x_1, A_1). Define

$$F = \{x \in M \mid f_1(x) = \ldots = f_n(x) = 0\}.$$

An open neighborhood Y of (x_1, A_1) in $U_1 \times W_1$ exists such that $\emptyset \neq g^{-1}(0) \cap Y \subseteq F \times G$. Hence

$$\dim F \times G \geqq n^2 + m - .1$$

Now, $\dim F \leqq m-2$, because $f_1,\ldots,f_n$ are coprime at every point of M. Hence $\dim F \times G \leqq m-2+n^2$. This is a contradiction. Hence $q = 1$. Similar $p = 1$ is proven. Therefore $g\, u_\mu = H_\mu$ and $g\, u_\nu = H_\nu$ and

$$g\, u_{\mu a_{\mu\rho}} + g_{a_{\mu\rho}} u_\mu = f_\rho$$

on $U_1 \times W_1$. Also g does not divide all $f_1,\ldots,f_n$ at (x_1, A_1).

Hence g does not divide $g_{a_{\mu\rho}}$ for at least one ρ. Take this ρ. Then

$$g\, u_{\nu a_{\mu\rho}} + g_{a_{\mu\rho}}\, u_{\nu} = 0$$

because $\mu \neq \nu$. Because g is prime at (x_1, A_1) and does not divide $g_{a_{\mu\rho}}$ at (x_1, A_1), it follows that g divides u_ν at (x_1, A_1) which contradicts the original choice of u_ν. Therefore H_μ and H_ν are coprime at every point of $M \times G$ if $\mu \neq \nu$.

Again, take a fixed pair (μ, ν) with $\mu \neq \nu$. Then $T = T_{\mu\nu} = H_{\mu\nu}^{-1}(0)$ is an analytic subset of $M \times G$ with

$$\dim T \leqq n^2 + m - 2$$

Let $\mathfrak{B}$ be the set of branches of T. Then $\dim B \leqq n^2+m-2$ if $B \in \mathfrak{B}$. Let $\varphi\colon T \to G$ be the projection $\varphi(x,A) = A$. Let $\varphi_B = \varphi|B$ be the restriction to $B \in \mathfrak{B}$. According to [2] Theorem 1.14

$$E_B = \{(x,A) \mid \operatorname{rank}_{(x,A)} \varphi_B \leqq n^2 - 1\}$$

is analytic in B. According to [2] Lemma 1.30, the image $E_B = \varphi(\tilde{E}_B)$ is almost thin of dimension $n^2 - 1$ in G. Hence $E = E_{\mu\nu} = \bigcup_{B \in T} E_B$ is almost thin of dimension $n^2 - 1$ in G. Particularly, E is nowhere dense in G.

Take any $(x,A) \in T$ with $A \in G - E$. Let C be a branch of $\varphi^{-1}(A) = \varphi^{-1}(\varphi(x,A))$ such that $(x,A) \in C$ and

$$\dim C = \dim_{(x,A)} \varphi^{-1}(A)$$

Then $C \subseteq B$ for at least one $B \in \mathcal{B}$. Hence $C \subseteq B \cap \varphi^{-1}(A) = \varphi_B^{-1}(A)$, and

$$n^2 = \operatorname{rank}_{(x,A)} \varphi_B = \dim B - \dim_{(x,A)} \varphi_B^{-1}(A)$$

$$\leqq n^2 + m - 2 - \dim_{(x,A)} \varphi_B^{-1}(A).$$

Hence

$$\dim C \leqq \dim_{(x,A)} \varphi_B^{-1}(A) \leqq m - 2$$

Hence $\dim_{(x,A)} \varphi^{-1}(\varphi(x,A)) \leqq m - 2$ for each $(x,A) \in T = T_{\mu\nu}$.

Define $H_{\mu A}: G \to \mathbb{C}$ by $H_{\mu A}(x) = H_\mu(x,A)$ and $H_{\mu\nu A} = (H_{\mu A}, H_{\nu A})$ by $H_{\mu\nu A}(x) = H_{\mu\nu}(x,A)$ for all $x \in M$.

Consider the point $x_0 \in M$ given in the assumption of the theorem. Since each $E_{\mu\nu}$ is nowhere dense, take $A \in G$ such that $A \notin E_{\mu\nu}$ for all pairs (μ,ν) with $\mu \neq \nu$. Now take any pair (μ,ν) with $\mu \neq \nu$. If $H_{\mu A}(x_0) \neq 0$ or $H_{\nu A}(x_0) \neq 0$, then $H_{\mu A}$ and $H_{\nu A}$ are coprime at x_0. Hence consider the case where $H_{\nu A}(x_0) = H_{\mu A}(x_0) = 0$. Then $(x,A) \in T = T_{\mu\nu}$. Now

$$H_{\mu\nu A}^{-1}(0) \times \{A\} = \{(x,A) \in M \mid (x,A) \in T_{\mu\nu}\} = \varphi^{-1}(A)$$

Because $\dim \varphi^{-1}(A) \leqq m - 2$, this implies

$$\dim_{x_0} H_{\mu\nu A}^{-1}(0) \leqq m - 2.$$

Therefore $H_{\mu A}$ and $H_{\nu A}$ are coprime at x_0. Define

$$h_\mu = H_{\mu A} = \sum_{\rho=1}^{n} a_{\mu\rho} f_\rho$$

Then $h_1, \ldots, h_n$ are pair wise coprime at A. q.e.d.

There ought to be a simpler proof around. Also, it may be of interest to know for which U.F.D. $\mathbb{C}$-algebras the Lemma holds.

Lemma 5.11. Assume (A1) - (A3). Let $\mathfrak{w} : M \to V$ be a holomorphic vector function such that $\dim \mathfrak{w}^{-1}(0) \leqq m - 2$. Let K be a compact subset of M. Then the integral

$$\int_K \frac{\mathfrak{w}^*(\upsilon_0) \wedge \chi}{|\mathfrak{w}|^2}$$

exists.

Proof. Let $\mathfrak{v} = (\mathfrak{v}_0, \ldots, \mathfrak{v}_n)$ be an orthonormal of V. Then

$$\mathfrak{w} = v_0 \mathfrak{v}_0 + \ldots + v_n \mathfrak{v}_n$$

where $v_0, \ldots, v_n$ are coprime at every point of M. It suffices to prove that $|\mathfrak{w}|^{-2} \mathfrak{w}^*(\upsilon_0) \wedge \chi$ is integrable over some neighborhood of each point $x_0 \in K$. A matrix $A = (a_{\mu\nu})$ with $\det A \neq 0$ exists such that

$$h_\mu = \sum_{\nu=0}^{n} a_{\mu\nu} v_\nu$$

are pair wise coprime at x_0. Hence they are coprime at every point of an open, connected neighborhood U of x_0. A base $\hat{\mathfrak{v}} = (\hat{\mathfrak{v}}_0, \ldots, \hat{\mathfrak{v}}_n)$ of V exists such that

$$\mathfrak{v}_\nu = \sum_{\mu=0}^{n} \hat{\mathfrak{v}}_\mu \, a_{\mu\nu}$$

Then

$$\mathfrak{w} = \sum_{\mu=0}^{n} v_\mu \, \mathfrak{v}_\mu = \sum_{\nu=0}^{n} h_\nu \, \hat{\mathfrak{v}}_\nu$$

Let $\| \ \|$ be the length of a hermitian product on V such that $\hat{\mathfrak{v}}$ is an orthonormal base for $\hat{\mathfrak{v}}$. Let $\hat{\upsilon}_0$ be the euclidean fundamental form for $\| \ \|$. A positive constant $L > 0$ exists such that

$$\frac{1}{L} |\mathfrak{z}| \leqq \|\mathfrak{z}\| \leqq L |\mathfrak{z}| \qquad \text{if } \mathfrak{z} \in V$$

$$\frac{1}{L} \upsilon_0 \leqq \hat{\upsilon}_0 \leqq L \upsilon_0$$

Hence

$$\frac{\mathfrak{w}^*(\upsilon_0)}{|\mathfrak{w}|^2} \leqq L^3 \frac{\mathfrak{w}^*(\hat{\upsilon}_0)}{\|\mathfrak{w}\|^2}$$

Hence it suffices to prove the integrability of $|\mathfrak{w}|^{-2} \mathfrak{w}^*(\hat{\upsilon}_0) \wedge \chi \geqq 0$ over some neighborhood of x_0. Let z_0, z_1 be the coordinate functions on $\mathbb{C}^2$ and define

$$\varphi = \frac{i}{2\pi} dz_0 \wedge d\bar{z}_0 \qquad \psi = \frac{1}{2\pi} dz_1 \wedge d\bar{z}_1$$

for $\nu = 0,1,\ldots,n$. Define

$$H_\nu = (h_0, h_\nu) : M \to \mathbb{C}^2$$

for $\nu = 1,\ldots,n$. Then

$$0 \leqq \frac{\mathfrak{w}^*(\hat{\upsilon}_0)}{|\mathfrak{w}|^2} \wedge \chi = \sum_{\nu=0}^{n} \frac{i}{2\pi} \frac{dh_\nu \wedge \overline{dh_\nu}}{|h_0|^2 + \ldots + |h_n|^2} \wedge \chi$$

$$\leqq \frac{i}{2\pi} \frac{H_1^*(\varphi)}{|H_1|^2} \wedge \chi + \sum_{\nu=1}^{n} \frac{i}{2\pi} \frac{H_\nu^*(\psi)}{|H_\nu|^2} \wedge \chi$$

If $H_\nu(x_0) \neq 0$ then the differential $|H_\nu|^{-2} H_\nu^*(\varphi) \wedge \chi$ respectively $|H_1|^{-2} H_1^*(\varphi) \wedge \chi$ is integrable over some neighborhood of x_0. Hence, assume $H_\nu(x_0) = 0$. Then $\dim_{x_0} H_\nu^{-1}(0) \leqq m - 2$ because h_0 and h_ν are coprime at x_0. Because H_ν is a map into $\mathbb{C}^2$, $\dim_x H_\nu^{-1} H_\nu(x) \geqq m - 2$ for all $x \in M$. Hence $\dim_{x_0} H_\nu^{-1}(0) = m - 2$. Therefore an open, connected neighborhood U_1 of x_0 in U exists such that $H_\nu : U_1 \to \mathbb{C}^2$ is $(m-2)$-fibering. Let U_2 be an open, connected neighborhood of x_0 such that $\overline{U}_2$ is compact and contained in U_1. According to [37] Proposition 1.7, the integral

$$\int_{U_2} \frac{i}{2\pi} \frac{H_\nu^*(\psi)}{|H_\nu|^2} \wedge \chi$$

exists. If $\nu = 1$, the same proposition implies the existence of

$$\int_{U_2} \frac{i}{2\pi} \frac{H_1^*(\varphi)}{|H_1|^2} \wedge \chi$$

this holds for $\nu = 1,\ldots,n$. Also U_2 can be chosen uniformly for all ν. Hence $\|\mathfrak{w}\|^{-2}\mathfrak{w}^*(\hat{\upsilon}_0) \wedge \chi$ is integrable over U_2.

q.e.d.

Proposition 5.12. Assume (A1), (A3) and (A4). Let ζ and ξ be continuous differential forms on $\mathbb{P}(V)$ and M respectively with $\deg \zeta = p$ and $\deg \xi = 2m - p$. Let K be any compact subset of M. Then $f^*(\zeta) \wedge \xi$ is integrable over K.

Proof. Let F be the closed graph of the meromorphic map f. Let $j: F \to M \times \mathbb{P}(V)$ be the inclusion. Let $\tilde{f}: F \to M$ and $\hat{f}: M \to \mathbb{P}(V)$ be the projections. Also let $\tilde{f}_0: M \times \mathbb{P}(V) \to M$ and $\hat{f}_0: M \times \mathbb{P}(V) \to \mathbb{P}(V)$ be the projections. Then $\tilde{f} = \tilde{f}_0 \circ j$ and $\hat{f} = \tilde{f} \circ j$. Define $A = M - I_f$ and $\tilde{A} = \tilde{f}^{-1}(A)$. Then $F - \tilde{A}$ has measure zero on F. Also $\tilde{f}: \tilde{A} \to A$ is biholomorphic. The set $\tilde{K} = \tilde{f}^{-1}(K)$ is compact. The differential form $\hat{f}_0^*(\zeta) \wedge \tilde{f}_0^*(\xi)$ is continuous on $M \times \mathbb{P}(V)$. According to Lelong [24].

$$j^*(\hat{f}_0^*(\zeta) \wedge \tilde{f}_0^*(\xi)) = \hat{f}^*(\zeta) \wedge \tilde{f}^*(\xi)$$

is integrable over $\tilde{K}$. Because $\tilde{f}: \tilde{K} \cap \tilde{A} \to K \cap A$ is the restriction of the biholomorphic map $\tilde{f}: \tilde{A} \to A$, the differential

$$(\tilde{f}^{-1})^*(\hat{f}^*(\zeta) \wedge \tilde{f}^*(\xi)) = f^*(\zeta) \wedge \chi$$

is integrable over $K - I_f$. Since I_f has measure zero, it is integrable over K.

q.e.d.

Especially the integrals

$$A_f(G) = \int_G f^*(\ddot{\omega}) \wedge \chi \geqq 0$$

$$\int_G \psi\, f^*(\ddot{\omega}) \wedge \chi \geqq 0$$

exist for each bump (G,g,ψ) on M.

Theorem 5.13. Assume (A1) - (A4). Let (G,g,ψ) be a bump on M. Then

$$T_f(G) = \int_G \psi\, f^*(\ddot{\omega}) \wedge \chi$$

Proof.[2] At first assume that an irreducible representation $\mathfrak{w}: U \to V$ of f on an open neighborhood U of $\overline{G}$ exists. Take $\varepsilon > 0$. Define $v_\varepsilon = \varepsilon + |\mathfrak{w}|^2 > 0$. Then

$$\int_{dG} \log v_\varepsilon\, d^\perp\psi \wedge \chi - \int_{dg} \log v_\varepsilon\, d^\perp\psi \wedge \chi - \int_{G-\overline{g}} \log v_\varepsilon\, dd^\perp\psi \wedge \chi$$

$$= \int_{G-\overline{g}} d \log v_\varepsilon \wedge d^\perp\psi \wedge \chi$$

$$= \int_{G-\overline{g}} d\, \psi \wedge d^\perp \log v_\varepsilon \wedge \chi$$

$$= -R \int_{dg} d^\perp \log v_\varepsilon \wedge \chi - \int_{G-\overline{g}} dd^\perp \log v_\varepsilon \wedge \chi$$

$$= \int_G \psi\, dd^c \log v_\varepsilon \wedge \chi$$

Observe $v_\varepsilon \to |\mathfrak{w}|^2$ monotonically for $\varepsilon \to 0$. Hence Theorem

2) This proof was presented at the Tulane conference. Later, B. Shiffman gave a similar proof in one of his talks at the conference. He uses currents and their regularizations.

5.9 implies

$$T_f(G) = \lim_{\varepsilon \to 0} \int_G \psi \frac{1}{4\pi} dd^c \log v_\varepsilon \wedge \chi$$

An easy computation shows

$$\frac{1}{4\pi} d^{\perp}d \log v_\varepsilon \wedge \chi = \frac{|\mathfrak{w}|^4}{v_\varepsilon^2} f^*(\omega_0) \wedge \chi + \frac{\varepsilon|\mathfrak{w}|^2}{v_\varepsilon^2} \frac{\mathfrak{w}^*(v)}{|\mathfrak{w}|^2} \wedge \chi$$

Here $|\mathfrak{w}|^2 \leqq v_\varepsilon$ and $\varepsilon < v_\varepsilon$ and $v_\varepsilon \to |\mathfrak{w}|^2$ for $\varepsilon \to 0$. Also $f^*(\ddot{\omega}_0) \wedge \chi$ and $|\mathfrak{w}|^{-2}\ \mathfrak{w}^*(\upsilon) \wedge \chi$ are integrable over $\overline{G}$ hence over G by Lemma 5.11 and Proposition 5.12. Hence $\varepsilon \to 0$ gives the desired result in this case.

Now, assume that $(G, \emptyset, \psi)$ is a simple bump. Take $a \in \overline{G}$. As in the proof of Theorem 5.3, an open neighborhood U(a) of a exists such that $(U(a) \cap G,\ 0,\ \lambda\psi)$ is a bump if $\lambda \geqq 0$ is a function of class C^∞ on M with compact support in U(a). Then

$$T_f(G, \lambda\psi) = \int_G \lambda\psi\ f^*(\omega_0) \wedge \chi$$

Finitely many $U(a_1), \ldots, U(a_p)$ cover $\overline{G}$. Take $\lambda_1, \ldots, \lambda_p$ as non-negative function of class C^∞ on M such that λ_μ has compact support in $U(a_\mu)$ and such that $\lambda_1 + \ldots + \lambda_p = 1$ in a neighborhood of $\overline{G}$. By the definition of T_f, it follows

$$T_f(G, \psi) = \sum_{\mu=1}^{p} T_f(G,\ {}_\mu\psi) = \int_G \psi\ f^*(\dot{\omega}_0) \wedge \chi$$

Now, consider the case of an arbitrary bump (G, g, ψ). Let $\tilde{\psi}$ be a function of class C^∞ on M such that $\tilde{\psi}(x) = \psi(x)$

if $x \in \overline{G}-g$ and such that $\tilde{\psi}|\overline{g} \geq R$. The definition of T_f implies

$$T_f(G,g,\psi) = T_f(G,\tilde{\psi}) - T_f(g,\tilde{\psi}-R)$$

$$= \int_G \tilde{\psi}\ f^*(\ddot{\omega}_0) \wedge \chi - \int_g (\tilde{\psi}-R)\ f^*(\ddot{\omega}_0) \wedge \chi$$

$$= \int_G \psi\ f^*(\ddot{\omega}_0) \wedge \chi$$

q.e.d.

Theorem 5.13 implies $T_f(G) \geq 0$. The second proof will be obtained in the next section as a side product of a more general procedure.

VI. AN INTEGRAL AVERAGE

Let V be a hermitian vector space of dimension n+1 with $n > 0$. Define $\tau: V \to \mathbb{R}$ by $\tau(\mathfrak{z}) = |\mathfrak{z}|^2$. Also define $\tau: V^* \to \mathbb{R}$ by $\tau(\mathfrak{z}) = |\mathfrak{z}|^2$. Let h be a measurable function almost everywhere defined on $\mathbb{P}(V^*)$. Assume that $h\ \ddot{\omega}^n$ is integrable over $\mathbb{P}(V^*)$. Define

$$I(h) = \int_{\mathbb{P}(V^*)} h\ \ddot{\omega}_0^n \tag{6.1}$$

It is well known (Weyl [48] Chapter III (2.8), Wu [51] Theorem 5.2, [39] Lemma 1.1, [34] Hilfsatz 1) that $h\ \ddot{\omega}_0^n$ is integrable over $\mathbb{P}(V^*)$ if and only if $e^{-\tau}(h \circ \mathbb{P})\ \upsilon^{n+1}$ is integrable over V^* and that

$$I(h) = \frac{1}{(n+1)!} \int_{V^*} e^{-\tau} (h \circ \mathbb{P})\, \upsilon^{n+1} \tag{6.2}$$

A continuous non-negative function w on $\mathbb{R}[0,1)$ shall be called a <u>weight function for n</u> if and only if a real number $q > 1$ exists such that the integral [3]

$$\int_0^1 (1-t)^{n-1}\, w(t)^q \, dt$$

exists. Then the following integrals exist

$$L(x) = L_w(x) = n \int_0^1 (1-t)^{n-1}\, w(xt)\, dt \tag{6.3}$$

for $0 \leqq x \leqq 1$ and

$$J(x) = J_w(x) = n(n-1) \int_0^1 (1-t)^{n-2}\, t\, w(xt)\, dt \quad \text{if } n \geqq 2 \tag{6.4}$$

$$J(x) = J_w(x) = w(x) \quad \text{if } n = 1 \tag{6.5}$$

for $0 \leqq x < 1$. Then

$$\int_0^1 J(x) dx = n(n-1) \int_0^1 (1-t)^{n-2} \int_0^t w(\tau)\, d\tau\, dt$$

$$= n \int_0^1 (1-t)^{n-1}\, w(t)\, dt = L(1)$$

Hence

---

3) For the following methods see Weyl [48] and Ahlfors [1].

$$L(1) = \int_0^1 J(x)\, dx \tag{6.6}$$

Also

$$J(x)\, x^n = n(n-1) \int_0^1 (x-t)^{n-2}\, tw(t)dt$$

Hence

$$w(x) = \frac{1}{x} \frac{1}{n!} \frac{d^{n-1}}{dx^{n-1}} (x^n J(x)) \tag{6.7}$$

for $0 < x < 1$.

The constant function $w \equiv 1$ is an example of a weight function. Then $L(x) \equiv 1$ and $J(x) \equiv 1$. If $0 < \lambda < 1$, another example of a weight function is given by

$$w_\lambda(t) = \frac{1}{t} \frac{1}{n!} \frac{d^{n-1}}{dt^{n-1}} \left(\frac{t^{n+1}}{(1-t)^\lambda}\right)$$

$$w_\lambda(t) = (n+1) \sum_{\nu=1}^{n-1} \binom{n-1}{\nu} \frac{\lambda(\lambda+1)\ldots(\lambda+\nu+1)}{(n-1-\nu)!} \frac{t^\nu}{(1-t)^{\lambda+\nu}} \geqq 0$$

A constant $c > 0$ exists, such c is independent of λ and such that

$$w_\lambda(t)\ (1-t)^{n-1-\lambda} \leqq c \quad \text{for all } 0 \leqq t < 1.$$

Take $q = \frac{1}{2} \frac{2n-1+\lambda}{n-1+\lambda} > 1$. Then

$$\int_0^1 w_\lambda^q(t)(1-t)^{n-1}\, dt \leqq \frac{2c^q}{1-\lambda} < +\infty \tag{6.8}$$

Therefore w_λ is a weight function. Here

$$J(x) = J_\lambda(x) = \frac{x}{(1-x)^\lambda} \tag{6.9}$$

$$L(1) = L_\lambda(1) = \frac{1}{(1-\lambda)(2-\lambda)} < \frac{1}{1-\lambda} \tag{6.10}$$

A diffeomorphism $\mu: \mathbb{P}(V) \to \mathbb{P}(V^*)$ is defined by $\mu \circ \mathbb{P} = \mathbb{P} \circ \tilde{\mu}$ with $(\tilde{\mu}(\mathfrak{b}); \mathfrak{x}) = (\mathfrak{x} | \mathfrak{b})$. If $b \in \mathbb{P}(V)$ and $x \in \mathbb{P}(V^*)$, then $||x;b|| = 1$ if and only if $x = \mu(b)$.

If $b \in \mathbb{P}(V)$ and if w is a weight function for n, define

$$\sigma_b(x) = w(||x;b||^2)$$

for all $x \in \mathbb{P}(V^*)$ with $x \neq \mu(b)$. If $w = w_\lambda$ with $\sigma_b = \sigma_{b,\lambda}$.

Lemma 6.1. Assume (A3). Let w be a weight function. Take $b \in \mathbb{P}(V)$. Then $I(\sigma_b) = L_w(1)$.
For the proof see Weyl [48] Chapter III Lemma 2.B and [36] 18.17.

Lemma 6.2. Assume (A3). Let w be a weight function. Take b and z in $\mathbb{P}(V)$. Define $h_z: \mathbb{P}(V^*) - \ddot{E}[z]$ by

$$h_z(x) = \log \frac{1}{||z;x||}$$

Then there exists a constant $c > 0$ such that $0 \leqq I(h_z\sigma_b) \leqq c$ for all $(z,b) \in \mathbb{P}(V) \times \mathbb{P}(V)$.

Moreover, if $0 < \lambda < 1$ and if $w = w_\lambda$, there exists a constant $c_n > 1$ depending only on n such that

$$0 \leqq I(H_z\ \sigma_b^{\lambda}) \leqq \frac{c_n}{(1-\lambda)^2}$$

for all $(z,b) \in \mathbb{P}(V) \times \mathbb{P}(V)$.

Proof. Take $q > 1$, such that $(1-t)^{n-1}\ w(t)^q$ is integrable over $\mathbb{R}[0,1]$. Define $p = \frac{q}{q-1} = 1$. Then $\frac{1}{p} + \frac{1}{q} = 1$. Hölders inequality implies

$$I(h_z\ \sigma_b) \leqq \frac{1}{2}\ I(2h_z)^p)^{\frac{1}{p}}\ I(\ {}^q_b)^{\frac{1}{q}} = c$$

By Lemma 6.1

$$I(2h_z)^p) = n \int_0^1 (1-t)^{n-1} (\log \tfrac{1}{t})^p\ dt = d_0$$

$$I(\sigma_b) = n \int_0^1 (1-t)^{n-1}\ w(t)^q\ dt = d_1$$

are constants. Hence $c = \frac{1}{2}\ d_0^{\frac{1}{p}}\ d_1^{\frac{1}{q}} \geqq 0$ is a constant.

Now assume $w = w_\lambda$. Take $q = \frac{1}{2}\ \frac{2n-1+\lambda}{n-1+\lambda} > 1$. According to (6.8), $d_1 \leqq \frac{2c_0^q}{(1-\lambda)}$, where c_0 is a constant independent of p.

If $x \geqq 1$, then $\log x \leqq 2p \log x^{\frac{1}{2p}} \leqq 2p\ x^{\frac{1}{2p}}$. Hence

$$d_0 \leqq n(2p)^p \int_0^1 (1-t)^{n-1} \frac{dt}{\sqrt{t}} = (2p)^p\ d_2$$

where d_2 depends on n only. Therefore

$$c = \tfrac{1}{2}(d_0)^{\frac{1}{p}}(d_1)^{\frac{1}{q}} \leqq p\ c_0 d_2^{\frac{1}{p}} \cdot (\tfrac{2}{1-\lambda})^{\frac{1}{q}} \leqq p\ c_0(d_2+1)\ \frac{2}{1-\lambda}$$

Here $p = \frac{q}{q-1} = \frac{2n-1+\lambda}{1-\lambda} \leqq \frac{2n}{1-\lambda}$. Hence

$$c \leqq \frac{c_n}{(1-\lambda)^2}$$

where c_n depends on n but not on λ;

q.e.d.

<u>Lemma 6.3</u>. Assume (A1). Define h_z as in Lemma 6.2. Then

$$I(h_z) = \frac{1}{2} \sum_{\nu=1}^{n} \frac{1}{\nu} \qquad \text{for all } z \in \mathbb{P}(V)$$

Proof.

$$I(h_z) = \frac{n}{2} \int_0^1 (1-t)^{n-1} \log \frac{1}{t}\, dt = \frac{1}{2} \sum_{\nu=1}^{n} \frac{1}{\nu}$$

q.e.d.

Take $b \in \mathbb{P}(V)$. A holomorphic map

$$\pi_b\colon \mathbb{P}(V) - \{b\} \to G_1(V)$$

is defined by $\pi_b(x) = x \wedge b$. The map is meromorphic on $\mathbb{P}(V)$. On $\mathbb{P}(V) - \{b\}$, the differential form

$$\Phi_b = \pi_b^*(\ddot{\omega}_1) \geqq 0$$

is realanalytic and has bidegree (1,1) with $d\Phi_b = 0$. According to [39] Lemma 3.3

$$\ddot{\omega}_0(x) - \Phi_b(x) = \frac{1}{2\pi} dd^{\perp} \log \|x:b\|$$

for $x \in \mathbb{P}(V) - \{b\}$.

According to [39] §5 p. 172-173, one and only one form ζ_b of bidegree (1,1) exists on $P(V) - \{b\}$ such that

$$\mathbb{P}^*(\zeta_b)(\mathfrak{z}) = \frac{i}{2\pi} \frac{|\mathfrak{z}|^4}{|\mathfrak{z} \wedge \mathfrak{w}|^2} \partial \frac{(\mathfrak{z}|b)}{|\mathfrak{z}|^2} \bar{\partial} \frac{(\mathfrak{z}|b)}{|\mathfrak{z}|^2}$$

where $\mathfrak{b} \in \mathbb{P}^{-1}(b)$. Then $\zeta_b \wedge \zeta_b = 0$ and

$$\ddot{\omega}_0(x) = \zeta_b(x) + ||x:b||^2 \Phi_b(x) \tag{6.11}$$

Take $b \in \mathbb{P}(V)$. Let w be a weight function. Define the form $\lambda_b \geqq 0$ on $\mathbb{P}(V^*) - \{\mu(b)\}$ by

$$\lambda_b(x) = g(||x;b||^2)\, \ddot{\omega}_0^n(x)$$

Recall, that

$$F = \{(x,y) \in \mathbb{P}(V) \times \mathbb{P}(V^*) \mid x \in \ddot{E}[y]\}$$

is a smooth, connected, compact complex submanifold $\mathbb{P}(V) \times \mathbb{P}(V^*)$ and that the projections $\pi: F \to \mathbb{P}(V)$ and $\psi: F \to \mathbb{P}(V^*)$ are regular, surjective, holomorphic maps of fiber dimension n-1.

Proposition 6.4. Assume (A3). Let w be a weight function for n. Take $(b,z) \in \mathbb{P}(V) \times \mathbb{P}(V)$ with $b \neq z$. Then the fiber integral $\tau_*\pi^*(\lambda_b)(z)$ exists. If $n \geqq 2$, then

$$\pi_*\psi^*(\lambda_b)(z) = L(||z:b||^2)\, ||z:b||^2 \Phi_b(z) + J(||z:b||^2)\, \zeta_b(z)$$

If n = 1, then

$$\pi_* \psi^*(\lambda_b)(z) = g(||z:b||^2)\ \zeta_b(z)$$

Here $\ddot{\omega}_0 = \zeta_b$.

Proof. At first, assume $n \geqq 2$. An orthonormal base $\mathfrak{w} = (\mathfrak{w}_0, \mathfrak{w}_1, \ldots, \mathfrak{w}_n)$ of V exists such that $\mathbb{P}(\mathfrak{w}_0) = z$ and $\mathbb{P}(\mathfrak{b}) = b$ with $\mathfrak{b} = b_0 \mathfrak{w}_0 + b_1 \mathfrak{w}_1$. Here $|\mathfrak{b}|^2 = |b_0|^2 + |b_1|^2 = 1$ can be assumed. Because $b \neq z$, the coordinate b_1 is not zero.

Consider the commutative diagram

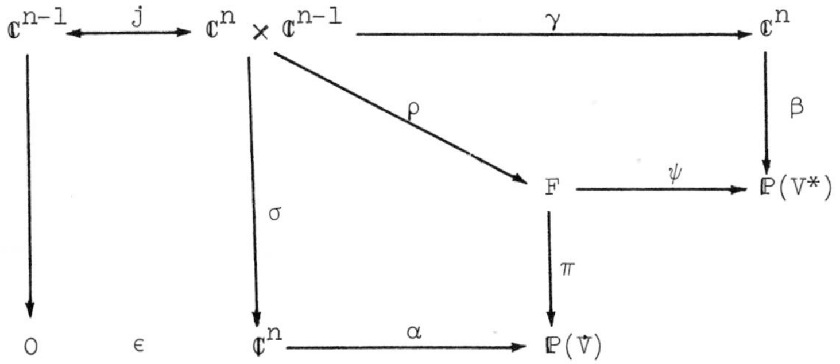

Here π, ψ, σ are projections, $j(y) = (0,y)$ for $y \in \mathbb{C}^{n-1}$. If $x = (x_1, \ldots, x_n) \in \mathbb{C}^n$, define

$$\alpha(x) = \mathbb{P}(\mathfrak{w}_0 + x_1 \mathfrak{w}_1 + \ldots + x_n \mathfrak{w}_n)$$

Let $\varepsilon_0, \ldots, \varepsilon_n$ be the dual base of $\mathfrak{w}$. If $y = (y_0, y_2, \ldots, y_n) \in \mathbb{C}^n$ define

$$\beta(y) = \mathbb{P}(y_0\varepsilon_0 + \varepsilon_1 + y_2\varepsilon_2 + \ldots + y_n\varepsilon_n)$$

If $x = (x_1, \ldots, x_n) \in \mathbb{C}^n$ and $u = (u_2, \ldots, u_n) \in \mathbb{C}^{n-1}$ define

$$h(x,u) = -x_1 - x_2u_2 - \ldots - x_nu_n$$

$$\gamma(x,u) = (h(x,u), u_2, \ldots, u_n)$$

Then $\alpha(\mathbb{C}^n) = U_\alpha$ and $\beta(\mathbb{C}^n) = U_\beta$ are open subsets of $\mathbb{P}(V)$ and $\mathbb{P}(V^*)$ respectively such that $\alpha\colon \mathbb{C}^n \to U_\alpha$ and $\beta : \mathbb{C}^n \to U_\beta$ are biholomorphic. Hence $\alpha \times \beta\colon \mathbb{C}^n \times \mathbb{C}^n \to U_\alpha \times U_\beta$ biholomorphically. Moreover

$$(\alpha \times \beta)^{-1}(F) = \{(x,y) \mid y_0 + x_1 + y_2x_2 + \ldots + y_nx_n = 0\}$$

$$= \{(x, \gamma(x,u)) \mid x \in \mathbb{C}^n, u \in \mathbb{C}^{n-1}\}.$$

Define

$$\rho(x,u) = (\alpha(x), \gamma(x,u)) \quad \text{for } (x,u) \in \mathbb{C}^n \times \mathbb{C}^{n-1}.$$

The maps in the diagram are defined. The diagram commutes. Observe $\alpha(0) = z$. The map

$$\rho \circ j\colon \mathbb{C}^{n-1} \to \pi^{-1}(z) \cap U_\beta$$

is biholomorphic. The complement $\pi^{-1}(z) - U_\beta$ is thin analytic on $\tau^{-1}(z)$. Hence, the fiber integral at z can be computed by the parameter representation j.

If $y = (y_0, y_2, \ldots, y_n)$, define $q = |y|^2$. Then

$$\beta^*(\ddot{\omega}_0) = \frac{i}{2\pi}\,\partial\bar{\partial} \log (1+q)$$

$$= \frac{i}{2\pi}\,[(1+q)^{-1}\,\partial\bar{\partial}q - (1+q)^{-2}\,\partial q \wedge \bar{\partial}q]$$

$$\beta^*(\ddot{\omega}_0^n) = (\tfrac{i}{2\pi})^n (1+q)^{-n-1}\,[(1+q)\,\partial\bar{\partial}q - n\,\partial q \wedge \bar{\partial}q] \wedge (\partial\bar{\partial}q)^{n-1}$$

$$0 = (\partial\bar{\partial} \log q)^n = q^{-n-1}\,(\partial\bar{\partial}q)^{n-1}\,[q\partial\bar{\partial}q - n\partial q \wedge \bar{\partial}q]$$

Therefore

$$\beta^*(\ddot{\omega}_0^n) = (\tfrac{i}{2\pi})^n\,(1 + |y|^2)^{-n-1}\,(\partial\bar{\partial}|y|^2)^n$$

Hence, at $(x,u) \in \mathbb{C}^n \times \mathbb{C}^{n-1}$

$$\rho^*\psi^*(\ddot{\omega}_0^n) = \gamma^*\beta^*(\ddot{\omega}_0^n)$$

$$= (\tfrac{i}{2\pi})^n (1 + |h|^2 + |u|^2)^{-n-1}\,[\partial\bar{\partial}(|h|^2 + |u|^2)]^n$$

$$= (\tfrac{i}{2\pi})^n (1 + |h|^2 + |u|^2)^{-n-1}\,n\,dh \wedge \overline{dh} \wedge (\partial\bar{\partial}|u|^2)^{n-1}$$

Define $u = (u_2, \ldots, u_n)$ and $u_1 = 1$. Then

$$dh \wedge d\bar{h} \wedge (\partial\bar{\partial}|u|^2)^n = \sum_{\mu,\nu=1}^{n} u_\mu \bar{u}_\nu\, dx_\mu \wedge d\bar{x}_\nu \wedge (\partial\bar{\partial}|u|^2)^{n-1}$$

Therefore

$$(\alpha^* \circ \pi_* \circ \psi^*)(\lambda_b)(0) = (\sigma_* \circ \gamma^* \circ \beta^*)(\lambda_b)(0) =$$

$$= \frac{i}{2\pi} \sum_{\mu,\nu=1}^{n} A_{\mu\nu}\, dx_\mu \wedge d\overline{x}_\nu$$

with

$$A_{\mu\nu} = n \int_{\mathbb{C}^{n-1}} w(||\beta \circ \gamma \circ j(u);b||^2)\, u_\mu \overline{u}_\nu (1+ |u|^2)^{-n-1}\, \upsilon^{n-1}$$

Here $\upsilon^{n-1} = (\frac{i}{2\pi})^{n-1}(\partial\overline{\partial}|u|^2)^{n-1}$ is the euclidean volume element. Also $h \circ j(u) = 0$ was used. Here

$$\beta \circ \gamma \circ j(u) = \beta \circ \gamma(0,u) = \beta(0,u) = \mathbb{P}(\varepsilon_1 + u_2\varepsilon_2 + \ldots + u_n\varepsilon_n).$$

$$b = \mathbb{P}(b_0 \mathfrak{w}_0 + b_1 \mathfrak{w}_1)$$

Hence

$$||\beta \circ \gamma \circ j(u);b||^2 = (1 + |u|^2)^{-1} |b_1|^2$$

$$||z:b||^2 = |\mathfrak{w}_0 \wedge (b_0 \mathfrak{w}_0 + b_1 \mathfrak{w}_1)|^2 = |b_1|^2$$

Hence

$$A_{\mu\nu} = n \int_{\mathbb{C}^{n-1}} w(\frac{||z:b||^2}{1+|u|^2})\, u_\mu \overline{u}_\nu\, (1 + |u|^2)^{-n-1}\, \upsilon^{n-1}$$

If $1 \leqq \nu < \mu \leqq n$ the substitution $u_\mu \to -u_\mu$ shows that $A_{\mu\nu} = 0$. If $1 \leqq \mu < \nu \leqq n$, then $A_{\mu\nu} = \overline{A}_{\nu\mu} = 0$. Observe $u_1 = 1$. Hence

$$A_{11} = n \int_{C^{n-1}} w\left(\frac{||z:b||^2}{1+|u|^2}\right) (1 + |u|^2)^{-n-1} \upsilon^{n-1}$$

Let Φ be the volume of the unit sphere in $\mathbb{C}^{n-1}$, then

$$1 = \int_{\mathbb{C}^{n-1}(1)} \upsilon^{n-1} = \int_0^1 r^{2n-3} \, dr \, \Phi = \frac{1}{2n-2} \Phi$$

Hence $\Phi = 2n - 2$. Therefore

$$A_{11} = n \int_0^{\infty} w\left(\frac{||z:b||^2}{1+r^2}\right) r^{2n-3} \, dr \, (2n-2)$$

Substitute $t = \frac{1}{1+r^2}$. Then

$$A_{11} = n(n-1) \int_0^1 w(t||z:b||^2)(1-t)^{n-2} \, tdt$$

$$= J(||z:b||^2)$$

If $\mu > 1$, then $A_{\mu\mu}$ does not depend on μ. Hence

$$A_{\mu\mu} = \frac{n}{n-1} \int_{\mathbb{C}^{n-1}} w\left(\frac{||z:b||^2}{1+|y|^2}\right) |y|^2 (1+|y|^2)^{-n-1} \upsilon^{n-1}$$

$$= 2n \int_0^{\infty} w\left(\frac{||z:b||^2}{1+r^2}\right) r^{2n-1} (1 + r^2)^{-n-1} \, dr$$

$$= n \int_0^1 w(||z:b||^2 t)(1-t)^{n-1} \, dt$$

$$= L(||z:b||^2)$$

Therefore

$$\alpha^*(\pi_*\psi^*(\lambda_b))(0) = J(||z:b||^2)\,\frac{i}{2\pi}\,dx_1 \wedge d\bar{x}_1 +$$

$$+ L(||z:b||^2)\,\frac{i}{2\pi}\sum_{\mu=2}^{n} dx_\mu \wedge d\bar{x}_\mu$$

For $x = (x_1,\ldots,x_n) \in \mathbb{C}^n$, define $\mathfrak{x} = \mathfrak{v}_0 + x_1\mathfrak{v}_1 + \ldots + x_n\mathfrak{v}_n$. Then

$$\alpha^*(\zeta_b)(0) = \frac{i}{2\pi}\,\frac{|\mathfrak{x}|^4}{|\mathfrak{b}\wedge\mathfrak{x}|^2}\,\partial\frac{(\mathfrak{x}|\mathfrak{b})}{|\mathfrak{x}|^2} \wedge \bar{\partial}\,\frac{(\mathfrak{b}|\mathfrak{x})}{|\mathfrak{x}|^2}\,\Big|_{x=0}$$

$$= \frac{i}{2\pi}\,\frac{1}{|b_1|^2}\,\bar{b}_1\,dx \wedge \quad b_1 d\bar{x}_1 = \frac{i}{2\pi}\,dx_1 \wedge d\bar{x}_1$$

$$\alpha^*(\ddot{\omega}_0)(0) = \frac{i}{2\pi}\,|\mathfrak{x}|^{-4}(|\mathfrak{x}|^2\,\partial\bar{\partial}|\mathfrak{x}|^2 - \partial|\mathfrak{x}|^2 \wedge \bar{\partial}|\mathfrak{x}|^2)\,|_{x=0}$$

$$= \frac{i}{2\pi}\,\partial\bar{\partial}|\mathfrak{x}|^2 = \frac{i}{2\pi}\sum_{\mu=1}^{n} dx_\mu \wedge d\bar{x}_\mu$$

Hence

$$\alpha^*(||z:b||^2\,\Phi_b(z))(0) = \alpha^*(\ddot{\omega}_0)(0) - \alpha^*(\zeta_b)(0) = \sum_{\mu=2}^{n}\frac{i}{2\pi}dx_\mu \wedge d\bar{x}_\mu$$

Therefore

$$\pi_*\psi^*(\lambda_b)(z) = J(||z:b||^2)\,\zeta_b(z) + L(||z:b||^2)\,||z:b||^2\,\Phi_b(z)$$

which proves the case $n > 1$.

Consider the case $n = 1$. Then π and τ are biholomorphic maps. Let $\mathfrak{v} = (\mathfrak{v}_0, \mathfrak{v}_1)$ be an orthonormal base of V and let $\varepsilon = (\varepsilon_0, \varepsilon_1)$ be the dual base. A linear isomorphism $D_{\mathfrak{v}} : V \to V^*$ is defined such that

$$(\mathfrak{x}, D_{\mathfrak{w}} \mathfrak{y})\, \mathfrak{w}_0 \wedge \mathfrak{w}_1 = \mathfrak{x} \wedge \mathfrak{y}$$

for all $\mathfrak{x}$ and $\mathfrak{y}$ in V. Hence a biholomorphic map $\hat{D}_{\mathfrak{w}} : \mathbb{P}(V) \to \mathbb{P}(V^*)$ exists uniquely, such that $\hat{D}_{\mathfrak{w}} \circ \mathbb{P} = \mathbb{P} \circ D_{\mathfrak{w}}$. Now $\hat{D}_{\mathfrak{w}} \circ \pi = \psi$ is claimed.

Take any $\mathfrak{x} \in V$, then $\mathfrak{x} = x_0 \mathfrak{w}_0 + x_1 \mathfrak{w}_1$ and

$$(\mathfrak{x}, D_{\mathfrak{w}} \mathfrak{w}_0)\, \mathfrak{w}_0 \wedge \mathfrak{w}_1 = \mathfrak{x} \wedge \mathfrak{w}_0 = -x_1 \mathfrak{w}_0 \wedge \mathfrak{w}_1$$

$$(\mathfrak{x}, D_{\mathfrak{w}} \mathfrak{w}_1)\, \mathfrak{w}_0 \wedge \mathfrak{w}_1 = \mathfrak{x} \wedge \mathfrak{w}_1 = x_0 \mathfrak{w}_0 \wedge \mathfrak{w}_1$$

Therefore $(\mathfrak{x}, D_{\mathfrak{w}} \mathfrak{w}_0) = -\varepsilon_1(\mathfrak{x})$ and $(\mathfrak{x}, D_{\mathfrak{w}} \mathfrak{w}_1) = \varepsilon_0(\mathfrak{x})$ for all $\mathfrak{x} \in V$ which implies

$$D_{\mathfrak{w}} \mathfrak{w}_0 = -\varepsilon_1 \qquad D_{\mathfrak{w}} \mathfrak{w}_1 = \varepsilon_0$$

Take $(x,y) \in F$. Take $\mathfrak{x} = x_0 \mathfrak{w}_0 + x_1 \mathfrak{w}_1 \in \mathbb{P}^{-1}(x)$ and $\mathfrak{y} = y_0 \mathfrak{w}_0 + y_1 \mathfrak{w}_1 \in \mathbb{P}^{-1}(y)$. Then $x_0 y_0 + x_1 y_1 = 0$. Because $\mathfrak{x} \neq 0 \neq \mathfrak{y}$ a number $\lambda \neq 0$ exists such that $x_1 = \lambda y_0$ and $x_0 = -\lambda y_1$. Therefore

$$D_{\mathfrak{w}}(\mathfrak{x}) = -x_0 \varepsilon_1 + x_1 \varepsilon_0 = \lambda(y_1 \varepsilon_1 + y_0 \varepsilon_0) = \lambda \mathfrak{y}$$

Hence

$$\hat{D}_{\mathfrak{w}}(\pi(x,y)) = \hat{D}_{\mathfrak{w}}(x) = \hat{D}_{\mathfrak{w}}(\mathbb{P}(\mathfrak{x})) = \mathbb{P}(D_{\mathfrak{w}}(\mathfrak{x})) = \mathbb{P}(\lambda \mathfrak{y}) =$$

$$= \mathbb{P}(\mathfrak{y}) = y = \psi(x,y).$$

Therefore $\hat{D}_{\mathfrak{v}} \circ \pi = \psi$. Observe that $D_{\mathfrak{v}}$ is an isometry $\tau \circ D_{\mathfrak{v}} = \tau$. Hence

$$D_{\mathfrak{v}}^*(\omega) = \frac{1}{4\pi} dd^{\perp} \log \tau \circ D_{\mathfrak{v}} = \frac{1}{4\pi} dd^{\perp} \log \tau = \omega$$

Hence $\hat{D}^*(\ddot{\omega}_0) = \ddot{\omega}_0$. Observe $||\hat{D}_{\mathfrak{v}} z;b|| = ||z:b||$. Therefore

$$\pi_*\psi^*(\lambda_b)(z) = (\psi \circ \pi^{-1})^*(\lambda_b)(z) = \hat{D}_{\mathfrak{v}}^*(\lambda_b)(z)$$

$$= ||\hat{D}_{\mathfrak{v}}(z);b||^2 \hat{D}_{\mathfrak{v}}^*(\ddot{\omega}_0)$$

$$= ||z:b||^2 \ddot{\omega}_0(z)$$

It remains to be proven that $\ddot{\omega}_0 = \zeta_b$ if $n = 1$. Take the orthonormal base $\mathfrak{v}_0, \mathfrak{v}_1$ such that $\mathbb{P}(\mathfrak{v}_0) = z$ and $\mathbb{P}(b_0 \mathfrak{v}_0 + b_1 \mathfrak{v}_1) = b$. Define $\alpha: \mathbb{C} \to \mathbb{P}(V)$ by $\alpha(x) = \mathbb{P}(\mathfrak{v}_0 + x \mathfrak{v}_1)$. Then $\alpha: \mathbb{C} \to \alpha(\mathbb{C})$ is biholomorphic and $\alpha(\mathbb{C})$ is an open neighborhood of z. As in the case $n \geqq 2$,

$$\alpha^*(\zeta_b)(0) = \frac{i}{2\pi} dx \wedge d\bar{x} = \alpha^*(\ddot{\omega}_0)(0)$$

Hence $\zeta_b(z) = \ddot{\omega}_0(z)$ for all $z \neq b$.

q.e.d.

<u>Proposition 6.5</u>. Assume (A1),(A3) and (A4). Suppose that f is generic. Let w be a weight function for n. Let K be a compact subset of M. Let $\eta \geqq 0$ be a continuous differentiable form of bidegree $(m-1, m-1)$ on M. Take $b \in \mathbb{P}(V^*)$. For each $y \in \mathbb{P}(V^*)$, define

$$n(K,\eta,y) = \int_{K \cap \gamma_f(y)} \delta_f^y \, \eta$$

With $+\infty$ as a value of an integral permitted, the following identity holds

$$\int_{\mathbb{P}(V^*)} n(K,\eta,y)\, w(||y:b||^2)\, \ddot{\omega}^n(y) =$$

$$= \int_K L(||f:b||^2)\, ||f:b||^2\, f^*(\Phi_b) \wedge \eta$$

$$+ \int_K J(||f:b||^2)\, f^*(\zeta_b) \wedge \eta$$

If $w \equiv 1$, then

$$0 \leqq \int_{\mathbb{P}(V^*)} n(K,\eta,y)\, \ddot{\omega}_0^n(y) = \int_K f^*(\ddot{\omega}_0) \wedge \eta < +\infty$$

Proof. Define $M_0 = M - I_f$ and $K_0 = K - I_f$. Then $f\colon M_0 \to \mathbb{P}(V)$ is holomorphic. According to [39] Lemma 2.2, the set

$$H = \{(x,y) \in M_0 \times \mathbb{P}(V^*) \mid f(x) \in \ddot{E}[y]\}$$

is a closed, smooth complex submanifold of dimension $m+n-1$ of $M_0 \times \mathbb{P}(V)$. The projection $\rho\colon H \to M_0$ is a regular, proper holomorphic map of pure fiber dimension $n-1$. Define $h\colon H \to F$ by $h(x,y) = (f(x),y)$. Then h is holomorphic with $\pi \circ h = f \circ \rho$. Moreover $\hat{h} = \psi \circ h\colon H \to \mathbb{P}(V^*)$ is the projection. For $y \in \mathbb{P}(V^*)$

$$\hat{h}^{-1}(y) = f^{-1}(\ddot{E}[y]) \cap M_0 \times \{y\}.$$

Hence

$$\rho_y = \rho: \hat{h}^{-1}(y) \to f^{-1}(\ddot{E}[y]) \cap M_0 = \gamma_f(y) \cap M_0$$

is biholomorphic. If $x \in M_0$, then

$$h_x = h: \rho^{-1}(x) \to \pi^{-1}(f(x))$$

is biholomorphic.

Because f is generic, $\gamma_f(y) = \emptyset$ or pure (m-1)-dimensional. Hence $\hat{h}$ has pure fiber dimension m-1 and is open. The set

$$S = \{(x,y) \in H \mid \hat{h} \text{ not regular at } (x,y)\}$$

is thin analytic in H and $S' = \hat{h}(S)$ has measure zero. The map $\hat{h}: H - S \to \mathbb{P}(V^*)$ is regular. By [39] Lemma 2.5 $\delta_f^y(x) = 1$ if $(x,y) \in H - S$. Define $\tilde{K} \doteq \rho^{-1}(K \cap M_0)$. Define $\tilde{K}_y = \hat{h}^{-1}(y) \cap \tilde{K}$

$$\begin{aligned}
\int_{\tilde{K}} \hat{h}^*(\lambda_b) \wedge \rho^*(\eta)) &= \int_{\mathbb{P}(V^*)} \int_{\tilde{K}_y} \rho^*(\eta)\ \lambda_b(y) \\
&= \int_{\mathbb{P}(V^*)} n(K \cap M_0, \eta, y)\ \lambda_b(y) \\
&= \int_{\mathbb{P}(V^*)} n(K, \eta, y)\ \lambda_b(y)
\end{aligned}$$

$$\int_{\tilde{K}} \hat{h}^*(\lambda_b) \wedge \rho^*(\eta) = \int_{\tilde{K}} h^*\psi^*(\lambda_b) \wedge \rho^*(\eta)$$

$$= \int_{K \cap M_0} \rho_*(h^*\psi^*(\lambda_b)) \wedge \eta$$

$$= \int_K f^*\pi_*\psi^*(\lambda_b) \wedge \eta$$

$$= \int_K L(||f:b||^2)||f:b||^2 f^*(\Phi_b) \wedge \eta$$

$$+ \int_K J(||f:b||^2) f^*(\zeta_b) \wedge \eta$$

If $w \equiv 1$, then $L(x) \equiv J(x) = 1$. Also

$$||f:b||^2 f^*(\Phi_b) + f^*(\zeta_b) = f^*(\ddot{\omega}_0)$$

by (6.11). Hence

$$\int_{\mathbb{P}(V^*)} n(K,\eta,y) \ddot{\omega}^n(y) = \int_K f^*(\ddot{\omega}_0) \wedge \eta$$

According to Proposition 5.12, this integral exists.

q.e.d.

Theorem 6.6. Assume (A1) - (A4). Let (G,g,ψ) be a bump on M. Then

$$T_f(G) = \int_G \psi\ f^*(\ddot{\omega}) \wedge \chi \tag{6.12}$$

If in addition f is generic, then

$$T_f(G) = \int_{\mathbb{P}(V^*)} N_f(G,y)\ \ddot{\omega}^n(y) \tag{6.13}$$

$$\int_{\mathbb{P}(V^*)} m_f(dG,y)\,\ddot{\omega}^n(y) = \sum_{\nu=1}^{n} \frac{1}{\nu}\frac{1}{4\pi}\int_{dG} d^{\perp}\psi \wedge \chi \qquad (6.14)$$

$$\int_{\mathbb{P}(V^*)} m_f^0(dg,y)\ddot{\omega}^n(y) = \Big(\sum_{\nu=1}^{n} \frac{1}{\nu}\Big) \frac{1}{4\pi}\int_{dG} d^{\perp}\psi \wedge \chi \qquad (6.15)$$

$$\int_{\mathbb{P}(V^*)} D_f(G,g)\ddot{\omega}^n(y) = \Big(\sum_{\nu=1}^{n} \frac{1}{\nu}\Big) \frac{1}{4\pi}\int_{G-\overline{g}} dd^{\perp}\psi \wedge \chi \qquad (6.16)$$

Proof. Assume that f is generic. Lemma 6.3 implies

$$I(m_f(dG,y)) = \frac{1}{4\pi}\int_{dG} I\Big(\log \frac{1}{\|f,y\|^2}\Big)\, d^{\perp}\psi \wedge \chi$$

$$= \sum_{\nu=1}^{n} \frac{1}{\nu}\frac{1}{4\pi}\int_{dG} d^{\perp}\psi \wedge \chi$$

which is (6.14). Similarly (6.15) and (6.16) are obtained. By Stokes Theorem

$$\frac{1}{4\pi}\int_{dG} d^{\perp}\psi \wedge \chi - \frac{1}{4\pi}\int_{dg} d^{\perp}\psi \wedge \chi = \frac{1}{4\pi}\int_{G-\overline{g}} dd^{\perp}\psi \wedge \chi$$

Hence the First Main Theorem implies (6.14). Now, Proposition 6.5 with $w \equiv 1$ implies (6.12). If f is not generic, then $f = j \circ f_0$ where $f_0: M \to \mathbb{P}(W)$ is a generic meromorphic map and where W is a linear subspace of V. Here $j: \mathbb{P}(W) \to \mathbb{P}(V)$ is the inclusion. The hermitian product on V restricts to a hermitian product on W such that $j^*(\ddot{\omega}_0)$ is the associated fundamental form of the induced Fubini Study Kaehler metric on $\mathbb{P}(W)$. Then

$$T_f(G) = T_{f_0}(G) = \int_G \psi\, f_0^* j_0^*(\ddot{\omega}_0) \wedge \chi = \int_G \psi\, f^*(\ddot{\omega}_0) \wedge \chi$$

q.e.d.

Since Theorem 5.13 was not used, Theorem 6.6 provides a second proof of (6.12). Define

$$\Phi(G) = \frac{1}{4\pi} \int_{dG} d^{\perp}\psi \wedge \chi =$$

$$= \frac{1}{4\pi} \int_{dg} d^{\perp}\psi \wedge \chi + \int_{G-\overline{g}} dd^{\perp}\psi \wedge \chi$$

By the first definition $\Phi(G) \geqq 0$. The sign of $dd^{\perp}\psi \wedge \chi$ may not be constant, but it may be necessary to estimate under the integral. Therefore, define

$$G_{+} = \{x \in G-\overline{g} \mid dd^{\perp}\psi \wedge \chi \geqq 0\}$$

$$\Phi^{+}(G) = \frac{1}{4\pi} \int_{dg} d^{\perp}\psi \wedge \chi + \int_{G_{+}} dd^{\perp}\psi \wedge \chi \geqq \Phi(G) \geqq 0$$

Theorem 6.7. Assume (A1) - (A4). Assume that f is generic. Let (G,g,ψ) be a bump on M. Then a constant $c > 0$ exists which depends on n and w only such that for each $b \in \mathbb{P}(V^*)$,

$$\int_{G} \psi\, L(||f:b||^2)\, ||f:b||^2\, f^*(\Phi_b) \wedge \chi$$

$$+ \int_{G} \psi\, J(||f:b||^2)\, f^*(\zeta_b) \wedge \chi$$

$$\leqq L(1)\, T_f(G) + c\, \Phi^{+}(G)$$

If $w = w_{\lambda}$, then there exists a constant $c_n > 2$, depending on n only, such that

$$(1-\lambda)^2 \ L(1) \leqq c_n - 1 \qquad (1-\lambda)c \leqq c_n - 1$$

Moreover, c_n can be taken as an increasing function of n.

Proof. Define

$$D_f^+(G,y) = \frac{1}{2\pi} \int_{G_+} \log \frac{1}{||f,y||} \ dd^\perp\psi \wedge \chi \geqq D_f(G,y)$$

The First Main Theorem implies

$$N_f(G,y) \leqq T_f(G) + m_f(dg,y) + D_f^+(G;y)$$

According to Lemma 6.2, a constant $c > 0$ exists such that

$$I \left(\log \frac{1}{||f,y||^2} \ \sigma_b(y)\right) \leqq c$$

for all $x \in M - I_f$ and $y \in \mathbb{P}(V^*)$. The constant c depends on n and w only. If $w = w_\lambda$, then $(1-\lambda)^2 c \leqq c_n - 1$ where $c_n > 2$ is a constant only depending on n. Obviously, c_n can be taken as to increase with n. Lemma 6.1 implies

$$I(\sigma_b(y)) = L(1)$$

If $w = w_\lambda$, then $L(1)(1-\lambda)^2 \leqq 1-\lambda \leqq 1 \leqq c_n - 1$ by (6.10). Therefore

$$I(N_f(G,y)\sigma_b(y)) \leqq L(1)T_f(G) + c \frac{1}{4\pi} \int_{dg} d^\perp\psi \wedge \chi + c \frac{1}{4\pi} \int_{G_+} dd^\perp\psi \wedge \chi$$

$$= L(1) \ T_f(G) + c \ \Phi^+(G)$$

Now, Proposition 6.5 with $\eta = \psi\chi$ proves the Theorem, q.e.d.

Let $\mu: V \to V^*$ be the antilinear map defined by $\mu(\mathfrak{z})(\mathfrak{x}) = (\mathfrak{x} | \mathfrak{z})$ for all $\mathfrak{x} \in V$ and $\mathfrak{z} \in V$. Then a diffeomorphism $\mu: \mathbb{P}(V) \to \mathbb{P}(V^*)$ is defined by $\mu \circ \mathbb{P} = \mathbb{P} \circ \mu$. Take $a \in \mathbb{P}(V^*)$. Define $b = \mu^{-1}(a)$. If $\alpha \in \mathbb{P}^{-1}(a)$, define $\mathfrak{b} = \mu^{-1}(\alpha)$. Then $\mathbb{P}(\mathfrak{b}) = b$. Also $\alpha(\mathfrak{x}) = \mu(\mathfrak{b})(\mathfrak{x}) = (\mathfrak{x} | \mathfrak{b})$ for all $\mathfrak{x} \in V$. Also $|\alpha| = |\mathfrak{b}|$. Hence

$$\mathbb{P}^*(\zeta_b)(\mathfrak{x}) = \frac{i}{2\pi} \frac{|\mathfrak{x}|^2 |\mathfrak{b}|^2}{|\mathfrak{x} \wedge \mathfrak{b}|^2} \frac{|\mathfrak{x}|^2}{|\alpha|^2} \partial\frac{\alpha(\mathfrak{x})}{|\mathfrak{x}|^2} \wedge \overline{\partial\frac{\alpha(\mathfrak{x})}{|\mathfrak{x}|^2}}$$

or

$$\mathbb{P}^*(||x:b||^2 \zeta_b)(\mathfrak{x}) = \frac{i}{2\pi} \frac{|\mathfrak{x}|^2}{|\alpha|^2} \partial \frac{\alpha(\mathfrak{x})}{|\mathfrak{x}|^2} \wedge \partial \frac{\alpha(\mathfrak{x})}{|\mathfrak{x}|^2}$$

Hence $||x:b||^2\, \zeta_b(x) \geqq 0$ is a form of class C^∞ and bidegree (1,1) on $\mathbb{P}(V^*)$. Hence a non-negative form ξ_a of class C^∞ and of bidegree (1,1) on $\mathbb{P}(V^*)$ is defined by

$$\xi_a(x) = ||x:b||^2\, \zeta_b(x) + ||x;a||^2\, \ddot{\omega}_0(x)$$

for all $x \in \mathbb{P}(V^*)$ where $b = \mu^{-1}(a)$.

If $f: M \to \mathbb{P}(V)$ is a meromorphic map, define

$$\mathfrak{A}_f = \{a \in \mathbb{P}(V^*) \mid f \text{ adapted to } a\}$$

According to Lemma 4.1, $\mathfrak{A}_f = \mathbb{P}(V^*) - \mathbb{P}(E)$, where $E \neq V^*$ is a linear subspace of V^*.

Theorem 6.8. Assume (A1) - (A4). Let (G,g,ψ) be a bump on M. Then there exists a constant $c_n > 2$, depending only on n and increasing with n such that for all λ with $0 < \lambda < 1$ and all $a \in \mathfrak{A}_f$ the following estimate holds

$$\int_G ||f;a||^{-2\lambda}\, \psi f^*(\xi_a) \wedge \chi \leqq \frac{c_n}{(1-\lambda)^2}\,[T_f(G) + \Phi^+(G)]$$

Proof. At first assume that f is generic. Then $\mathfrak{A}_f = \mathbb{P}(V^*)$. Take $0 < \lambda < 1$ and $a \in \mathbb{P}(V^*)$. Take w_λ as weight function for n. Define $b = \mu^{-1}(a)$. Then

$$||f:b||^2 + ||f;a||^2 = 1$$

Also $J(x) = x(1-x)^{-\lambda}$ by (6.9). Therefore Theorem 6.7 implies

$$\int_G \psi\, ||f;a||^{-2\lambda}\, ||f:b||^2\, f^*(\zeta_b) \wedge \chi$$

$$\leqq \frac{c_{n-1}}{(1-\lambda)^2}\,[T_f(G) + \Phi^+(G)]$$

Theorem 6.6 implies

$$\int_G \psi\, ||f;a||^{2-2\lambda}\, f^*(\ddot{\omega}_0) \wedge \chi \leqq \int_G \psi\, f^*(\ddot{\omega}_0) \wedge \chi = T_f(G)$$

Since $(1-\lambda)^{-2} \geqq 1$, addition proves the estimate if f is generic.

Now, consider the case where f is not generic. A linear subspace W of V with minimal dimension p+1 exists such that $f(M) \subseteq \mathbb{P}(W)$. Then W is unique. If $\mathfrak{w} : U \to V$ is any irreducible representation of f on an open subset U of M, then W is the linear hull of $\mathfrak{v}(U)$. If $p = 0$, then f is constant and $f^*(\ddot{\omega}_0) = 0 = f^*(\xi_a)$. Therefore the estimate is trivial. Hence $p > 0$ can be assumed. Because f is not generic, $p < n$. Let $j: W \to V$ and $\iota: \mathbb{P}(W) \to \mathbb{P}(V)$ be the inclusion maps. Then $\mathbb{P} \circ j = \iota \circ \mathbb{P}$. A generic meromorphic map $f_0: M \rightharpoonup \mathbb{P}(W)$ exists uniquely such that $\iota \circ f_0 = f$. The hermitian product on W is obtained as the restriction of the hermitian product on V. Hence j is an isometry and $T_{f_0}(G) = T_f(G)$.

Let $\hat{j}: V^* \to W^*$ be the adjoint of the linear map $j: W \to V$. If $\alpha \in V^*$, then $\hat{j}(\alpha) = \alpha \circ j$. Because j is an isometry, $\hat{j}$ is an orthonormal projection. Hence $|\alpha \circ j| \leqq |\alpha|$ for all $\alpha \in V^*$. Let E be the kernel of $\hat{j}$. Let $\mathfrak{v}: U \to V$ be an irreducible representation of f on an open connected subset U of M. Then $\mathfrak{v} = j \circ \mathfrak{v}_0$ where $\mathfrak{v}_0: U \to W$ is an irreducible representation of f_0 on U. Take $a \in \mathbb{P}(V^*)$. Take $\alpha \in \mathbb{P}^{-1}(a)$. If $a \in \mathbb{P}(E)$, then $\alpha \circ j \equiv 0$. Therefore $\alpha \circ \mathfrak{v} = \alpha \circ j \circ \mathfrak{v}_0 \equiv 0$. Hence $a \in \mathbb{P}(V) - \mathfrak{A}_f$. If $a \in \mathbb{P}(V) - \mathfrak{A}_f$, then $\alpha \circ \mathfrak{v} \equiv 0$. Therefore $\mathfrak{v}(U) \subseteq \alpha^{-1}(0)$. The linear hull of $\mathfrak{v}(U)$ is W. Hence $W \subseteq \alpha^{-1}(0)$ and $\alpha \circ j \equiv 0$. Therefore $\alpha \in E$ and $a \in \mathbb{P}(E)$. Hence

$$\mathfrak{A}_f = \mathbb{P}(V^*) - \mathbb{P}(E).$$

The linear map $\hat{j}: V^* \to W^*$ defines a holomorphic map

$$\hat{\iota}: \mathfrak{A}_f \to \mathbb{P}(W^*)$$

such that $\hat{\iota} \circ \mathbb{P} = \mathbb{P} \circ \hat{j}$. Take $a \in \mathfrak{A}_f$. Define $c = \hat{\iota}(a)$. Take $\alpha \in \mathbb{P}^{-1}(a)$. Then $\gamma = \hat{j}(\alpha) = \alpha \circ j \in W^* - \{0\}$ and $c = \mathbb{P}(\gamma)$. Also $|\gamma| = \rho|\alpha|$ with $0 < \rho \leqq 1$. Take $y \in \mathbb{P}(W)$. Define $x = \iota(y) \in \mathbb{P}(V)$. Take $\mathfrak{y} \in \mathbb{P}^{-1}(y)$. Then $\mathfrak{x} = j(\mathfrak{y}) \in \mathbb{P}(V)$ with $\mathbb{P}(\mathfrak{x}) = x$. Therefore

$$||x;a|| = \frac{|\alpha(\mathfrak{x})|}{|\mathfrak{x}||\alpha|} = \frac{|\alpha \circ j(\mathfrak{y})|}{|j(\mathfrak{y})||\alpha|} = \frac{|\gamma(\mathfrak{y})|}{|\mathfrak{y}||\gamma|}\rho = ||y;c||\rho$$

Also

$$\mathbb{P}^*(\iota^*(\xi_a))(\mathfrak{y}) = j^*(\mathbb{P}^*(\xi_a))(\mathfrak{y}) =$$

$$= \frac{i}{2\pi}\frac{|\mathfrak{x}|^2}{|\alpha|^2}\partial\frac{\alpha(\mathfrak{x})}{|\mathfrak{x}|^2} \wedge \overline{\partial\frac{\alpha(\mathfrak{x})}{|\mathfrak{x}|^2}} + ||x;a||^2\omega_0(\mathfrak{x})$$

$$= \rho^2\left(\frac{i}{2\pi}\frac{|\mathfrak{y}|^2}{|\gamma|^2}\partial\frac{\gamma(\mathfrak{y})}{|\mathfrak{y}|^2} \wedge \overline{\partial\frac{\gamma(\mathfrak{y})}{|\mathfrak{y}|^2}} + ||y;c||^2\omega_0(\mathfrak{y})\right)$$

$$= \rho^2\,\mathbb{P}^*(\xi_a)(\mathfrak{y}).$$

Hence $\iota^*(\xi_a) = \rho^2(\xi_a)$ and $f^*(\xi_a) = \rho^2 f_0^*(\xi_a)$. The first part of the proof implies

$$\int_G ||f;a||^{-2\lambda}\psi f^*(\xi_a) \wedge \chi =$$

$$= \int_G ||f_0;a||^{-2\lambda}\rho^{2-2\lambda}\psi f_0^*(\xi_a) \wedge \chi$$

$$\leqq \rho^{2-2\lambda}\frac{c_p}{(1-\lambda)^2}(T_{f_0}(G) + \Phi^+(G))$$

$$\leqq \frac{c_n}{(1-\lambda)^2}[T_f(G) + \Phi^+(G)]$$

because $2 - 2\lambda > 0$, because $0 < \rho \leqq 1$ and because $c_p \leqq c_n$;
q.e.d.

It is quite essential that Theorem 6.8 be established for non-generic maps to avoid awkward assumptions later on. For the final results to be achieved in this work, it would suffice to take $\lambda = \frac{1}{2}$. However, in view of later use for a proof of a defect relations, it seems to be advisable to study the dependence on λ for a while.

VII. ASSOCIATED MAPS

If $m = 1$, the last estimate leads to the defect relation using the associated maps. If $m > 1$, no natural definition of associated maps seems to be at hand without further assumptions. Reflecting on the one variable proof Ahlfors [1] and Weyl [48], it was possible to find a way to introduce associated maps in [36]. This method shall be used and improved here.

If X and Y are complex spaces. Let Hol (X;Y) be the set of all holomorphic maps of X into Y. Define Hol (X) = Hol(X,$\mathbb{C}$). Let M be a complex manifold of dimension $m > 0$. Let V be a complex vector space of dimension n+1 with $n \geqq 0$. Let $U \neq \emptyset$ be an open subset of M. Take a patch $\alpha = (\alpha_1, \ldots, \alpha_m) \in \mathfrak{T}_M$ with $U \cap U_\alpha \neq \emptyset$. Then a derivation

$$\frac{\partial}{\partial \alpha_\nu} : \text{Hol }(U,V) \to \text{Hol }(U \cap U_\alpha, V)$$

is defined by

$$\mathfrak{w}_{\alpha_\nu} = \frac{\partial}{\partial_{\alpha_\nu}} \mathfrak{w} = \left(\frac{\partial}{\partial z_\nu} (\mathfrak{w} \circ \alpha^{-1})\right) \circ \alpha$$

where $\frac{\partial}{\partial z_\nu}$ denotes the νth partial derivative on $\mathbb{C}^m$. The iterations are defined as usual.

A holomorphic differential form B of bidegree (m-1,0) on M is said to define a <u>contravariant differentiation</u> on M. If $U \neq \emptyset$ is open in M and if $\alpha \in \mathfrak{T}_M$ is a patch with $U \cap U_\alpha \neq \emptyset$, a derivation

$$D_{B,\alpha} = D_\alpha : \text{Hol}(U,V) \to \text{Hol}(U \cap U_\alpha, V)$$

is defined by

$$d\,\mathfrak{w} \wedge B = (D_{B\alpha}\mathfrak{w})\, d\alpha_1 \wedge \dots \wedge d\alpha_m$$

for each $\mathfrak{w} \in \text{Hol}(U,V)$. Write also

$$\mathfrak{w}' = \mathfrak{w}'_\alpha = D_\alpha \mathfrak{w} = D_{B\alpha}\mathfrak{w}.$$

The iteration is defined by induction

$$D^p_{B,\alpha}\mathfrak{w} = D_{B,\alpha}(D^{p-1}_{B,\alpha}\mathfrak{w}) \qquad \text{if } p > 1$$

Write also $D^0_{B,\alpha}\mathfrak{w} = \mathfrak{w}$ and $\mathfrak{w}^{(p)} = \mathfrak{w}^{(p)}_\alpha = D^p_{B,\alpha}\mathfrak{w}$. Holomorphic functions B_μ exist on U_α such that

$$B = \sum_{\mu=1}^{m} (-1)^{\mu-1} B_\mu \; d\alpha_1 \wedge \cdots \wedge d\alpha_{\mu-1} \wedge d\alpha_{\mu+1} \wedge \cdots \wedge d\alpha_m$$

Then

$$D_{B,\alpha}\mathfrak{w} = \mathfrak{w}'_\alpha = \sum_{\mu=1}^{m} B_\mu \; \mathfrak{w}_{\alpha_\mu}$$

Hence $\mathfrak{w}'_\alpha$ can be considered a directional derivative into the direction of the vector $(B_1,\ldots,B_m)$. However B does not define a vector field.

If U is open in M, if $\alpha \in \mathfrak{T}_M$ and if $U \cap U_\alpha \neq \emptyset$, then

$$D_\alpha(h\mathfrak{w}) = (D_\alpha h)\mathfrak{w} + hD_\alpha \mathfrak{w} \qquad \text{if } \begin{cases} h \in \text{Hol}\,(U) \\ \mathfrak{w} \in \text{Hol}\,(U,V) \end{cases}$$

$$D_\alpha(\mathfrak{w} + \mathfrak{w}\mathfrak{o}) = D_\alpha(\mathfrak{w}) + D_\alpha \mathfrak{w}\mathfrak{o} \qquad \text{if } \begin{cases} \mathfrak{w} \in \text{Hol}\,(U,V) \\ \mathfrak{w}\mathfrak{o} \in \text{Hol}\,(U,V) \end{cases}$$

$$D_\alpha(\mathfrak{w} \wedge \mathfrak{w}\mathfrak{o}) = D_\alpha(\mathfrak{w}) \wedge \mathfrak{w}\mathfrak{o} + \mathfrak{w} \wedge D_\alpha(\mathfrak{w}\mathfrak{o}) \quad \text{if } \begin{cases} \mathfrak{w} \in \text{Hol}(U,\textstyle\bigwedge_p V) \\ \mathfrak{w}\mathfrak{o} \in \text{Hol}(U,\textstyle\bigwedge_q V) \end{cases}$$

$$(\mathfrak{w}_\alpha^{(p)})_\alpha^{(q)} = \mathfrak{w}_\alpha^{(p+q)} \qquad \text{if } \mathfrak{w} \in \text{Hol}\,(U,V)$$

Let $\alpha \in \mathfrak{T}_M$ and $\beta \in \mathfrak{T}_M$ be patches with $U_\alpha \cap U_\beta \neq \emptyset$. Then

$$d\alpha_1 \wedge \cdots \wedge d\alpha_m = \Delta_{\alpha\beta} \, d\beta_1 \wedge \cdots \wedge d\beta_m$$

on $U_\alpha \cap U_\beta$. Here $\Delta_{\alpha\beta}$ is the Jacobian and $\Delta_{\alpha\beta}(x) \neq 0$ for all $x \in U_\alpha \cap U_\beta$. Of course, $\Delta_{\alpha\beta}$ is a holomorphic function. The family $\{\Delta_{\alpha\beta}\}$ is a holomorphic cocycle which defines the canonical line bundle K on M. These notations will be used consistantly. If $\mathfrak{w} \in \mathrm{Hol}\,(U,V)$ then

$$\mathfrak{w}'_\beta = \mathfrak{w}'_\alpha \, \Delta_{\alpha\beta} \qquad \text{on } U \cap U_\alpha \cap U_\beta \neq \emptyset$$

Let $V_M = M \times V$ be the trivial bundle, then $\mathfrak{w}'_\alpha$ can be viewed as the coordinate function of a holomorphic section $d\,\mathfrak{w} \wedge B$ on $K \otimes V_M$ over U. The higher derivatives cannot be interpreted this way. However, a similar fact holds for their exterior product

$$\mathfrak{w}_p = \mathfrak{w}_{p\alpha} = \mathfrak{w} \wedge \mathfrak{w}'_\alpha \wedge \cdots \wedge \mathfrak{w}^{(p)}_\alpha$$

Again this notation will be used consistantly throughout this paper.

Lemma 7.1. Assume (A1) and (A2). Let $U \neq \emptyset$ be open in M. Let $\alpha \in \mathfrak{T}_M$ and $\beta \in \mathfrak{T}_M$ be patches such that $U \cap U_\alpha \cap U_\beta \neq \emptyset$. Let $\mathfrak{w} : U \to V$ be holomorphic. Then holomorphic functions g_{pq} exist on $U \cap U_\alpha \cap U_\beta$ with $g_{pq} = \Delta^p_{\alpha\beta}$ such that

$$\mathfrak{w}^{(p)}_\beta = \sum_{q=1}^{p} g_{pq} \, \mathfrak{w}^{(q)}_\alpha \qquad \text{on } U \cap U_\alpha \cap U_\beta$$

Proof. The statement is true for $p = 1$. Assume it is correct for p. Then it shall be proved for $p + 1$.

$$\mathfrak{w}^{(p+1)}_\beta = \sum_{q=1}^{p} (g_{p,q})'_\beta \, \mathfrak{w}^{(q)}_\alpha + \sum_{q=1}^{p} g_{pq} \, \Delta_{\alpha\beta} \, \mathfrak{w}^{(q+1)}_\alpha$$

$$= \sum_{q=1}^{p+1} g_{p+1,q} \; \mathfrak{w}_{\alpha}^{(q)}$$

with $g_{p+1,p+1} = g_{pp} \, \Delta_{\alpha\beta} = \Delta_{\alpha\beta}^{p+1}$, q.e.d.

<u>Lemma 7.2</u>. Under the assumptions of Lemma 7.1

$$\mathfrak{w}_{p\alpha} = \mathfrak{w}_{p\beta} \, (\Delta_{\alpha\beta})^{\frac{p(p+1)}{2}}$$

Proof. The statement is correct for $p = 0,1$. Assume it is correct for $p - 1$. Then it shall be proved for p.

$$\mathfrak{w}_{p,\beta} = \mathfrak{w}_{p-1,\beta} \wedge \mathfrak{w}_{\beta}^{(p+1)} =$$

$$= \mathfrak{w} \wedge \mathfrak{w}'_{\alpha} \wedge \dots \wedge \mathfrak{w}_{\alpha}^{(p-1)} \wedge \Big(\sum_{q=1}^{p} g_{pq} \, \mathfrak{w}_{\alpha}^{(q)}\Big)(\Delta_{\alpha\beta})^{\frac{p(p-1)}{2}}$$

$$= \mathfrak{w} \wedge \mathfrak{w}'_{\alpha} \wedge \dots \wedge \mathfrak{w}_{\alpha}^{(p)} \; g_{pp} \, (\Delta_{\alpha\beta})^{\frac{p(p-1)}{2}}$$

$$= \mathfrak{w}_{p\alpha} \, (\Delta_{\alpha\beta})^{\frac{p(p+1)}{2}}$$

q.e.d.

<u>Lemma 7.3</u>. Assume (A1) and (A2). Let $U \neq \emptyset$ be open in M. Take $\alpha \in \mathfrak{T}_M$ with $U \cap U_\alpha \neq \emptyset$. Let $\mathfrak{w} : U \to V$ and $h: U \to \mathbb{C}$ be holomorphic. Define $\mathfrak{u} = h \, \mathfrak{w}$. Then

$$\mathfrak{u}_{p\alpha} = h^{p+1} \, \mathfrak{w}_{p\alpha}$$

Proof. The statement is correct for $p = 0$. Assume it is correct for p. Then it shall be proven for $p + 1$.

$$\mathfrak{w}_{p+1,\alpha} = \mathfrak{w}_{p,\alpha} \wedge (h\mathfrak{v})^{(p+1)}_{\alpha} =$$

$$= h^{p}\ \mathfrak{v}\wedge\mathfrak{v}'_{\alpha}\wedge\cdots\wedge\mathfrak{v}^{(p)}_{\alpha}\wedge\sum_{q=0}^{p+1}\binom{p+1}{q}h^{(p+1-q)}\,\mathfrak{v}^{(q)}_{\alpha}$$

$$= h^{p+1}\,\mathfrak{v}_{p+1,\alpha}$$

q.e.d.

Let f: $M \longrightarrow \mathbb{P}(V)$ be a meromorphic map. Then f is said to be <u>general of order p for B</u> if and only if a patch $\alpha \in \mathfrak{T}_M$ and an irreducible representation $\mathfrak{v}: U_\alpha \to V$ of f exist such that $\mathfrak{v}_{p\alpha} \not\equiv 0$ on U_α. The meromorphic map f is said to be <u>general for B</u> if and only if it is general of order n for B. If f is general of order p for B, and if $0 \leqq q \leqq p$ then f is general of order q for B.

<u>Lemma 7.4</u>. Assume (A1), (A3) and (A4). Assume that f is general of order B for p. Let $\alpha \in \mathfrak{T}_M$ be a patch and let $\mathfrak{v}: U_\alpha \to V$ be a representation of f on U_α. Then $\mathfrak{v}_{p,\alpha} \not\equiv 0$ on every, non-empty open subset of U_α.

Proof. Let $\mathfrak{A}$ be the set of all pairs $(\beta, \mathfrak{w})$ such that $\beta \in \mathfrak{T}_M$ and such that $\mathfrak{w}: U_\beta \to V$ is an irreducible representation of f on the open connected set U_β. Define

$$\mathfrak{A}_1 = \{(\beta,\mathfrak{w}) \in \mathfrak{A} \mid \mathfrak{w}_{p,\beta} \not\equiv 0\}$$

$$\mathfrak{A}_2 = \{(\beta,\mathfrak{w}) \in \mathfrak{A} \mid \mathfrak{w}_{p,\beta} \equiv 0\}.$$

$$M_\lambda = \bigcup_{(\beta,\mathfrak{w}) \in \mathfrak{A}_\lambda} U_\beta \qquad \text{for } \lambda = 1,2.$$

By assumption $\mathfrak{A}_1 \neq \emptyset$. Hence $M_1 \neq \emptyset$. Both sets M_1 and M_2 are open with $M = M_1 \cup M_2$. If $a \in M_1 \cap M_2$, then $a \in U_\beta \cap U_\gamma$ with $(\beta, \mathfrak{w}) \in \mathfrak{A}_1$ and $(\gamma, \mathfrak{z}) \in \mathfrak{A}_2$. Then $\mathfrak{z} = h\mathfrak{w}$ on $U_\beta \cap U_\gamma$ where $h\colon U_\beta \cap U_\gamma \to \mathbb{C} - \{0\}$ is holomorphic. Hence

$$0 \equiv \mathfrak{z}_{p,\gamma} = \mathfrak{w}_{p,\beta}\, h^{p+1}(\Delta_{\beta\gamma})^{\frac{p(p+1)}{2}} \not\equiv 0$$

on $U_\beta \cap U_\gamma \neq \emptyset$. This is a contradiction. Hence $M_1 \cap M_2 = \emptyset$. Since M is connected, $M_1 = M$ and $M_2 = \emptyset$. Hence $\mathfrak{A}_2 = \emptyset$.

Take $\alpha \in \mathfrak{T}_M$. Assume $\mathfrak{v}\colon U_\alpha \to V$ is a representation of f on U_α. Assume an open subset $U \neq \emptyset$ of U_α exists such that $\mathfrak{v}_{p,\alpha}|U \equiv 0$. An open, connected, subset $U_\beta \neq \emptyset$ and an irreducible representation $\mathfrak{w}\colon U_\beta \to V$ exist. Define $\beta = \alpha|U_\beta$. Then $\beta \in \mathfrak{T}_M$ and $(\beta, \mathfrak{w}) \in \mathfrak{A} = \mathfrak{A}_1$. A holomorphic function $h \not\equiv 0$ on U_β exists such that $\mathfrak{v} = h\mathfrak{w}$. Then $0 \equiv \mathfrak{v}_{p\alpha}|U_\beta = h^{p+1}\mathfrak{w}_{p\beta} \not\equiv 0$. This is a contradiction. Hence $\mathfrak{v}_{p,\alpha}|U \not\equiv 0$ if $U \not\equiv \emptyset$ is open in U_α,

q.e.d.

Let $f\colon M \to \mathbb{P}(V)$ be a meromorphic map, which is general of order p for B. Lemma 7.2 - 7.4 imply, that there exists one and only one meromorphic map

$$f_p\colon M \to \mathbb{P}(\textstyle\bigwedge_{p+1} V)$$

such that $\mathfrak{v}_{p,\alpha}$ is a representation of f_p on U_α whenever $\alpha \in \mathfrak{T}_M$ and $\mathfrak{v}\colon U_\alpha \to V$ is a representation of f on U_α. Observe, if $\mathfrak{v}$ is irreducible, $\mathfrak{v}_{p,\alpha}$ may not be irreducible. Observe

$$f_p(M) \subseteq G_p(V)$$

The meromorphic map f_p is called the pth associated map of f defined by B. Obviously, f_p depends on the choice of B if $p \geqq 1$. If $p = n$, then f_n exists but is a constant map $f_n: M \to \mathbb{P}(\mathbb{C})$, since $\mathbb{P}(\mathbb{C})$ consists of one point.

If $m = 1$, then B is a holomorphic function. A natural choice is $B \equiv 1$, which shall always be made in this paper, if $m = 1$. It is well known, that f is general for $B \equiv 1$ if and only if f is generic, i.e. if f(M) is not contained in any proper, projective subplane of $\mathbb{P}(V)$. In order not to operate in a vacuum, it is important to find reasonable conditions for f to be B-general for properly chosen B if $m > 1$. This shall be accomplished now.

Lemma 7.5. Let $G \neq \emptyset$ be an open, connected, simply connected subset of $\mathbb{C}$. Let V be a complex vector space. Let $\mathfrak{w}: G \to V$ be a holomorphic vector function. Assume $\mathfrak{w}' = g\mathfrak{w}$ where g is a holomorphic function on G. Then there exists a constant vector $\mathfrak{v} \in V$ and a holomorphic function h on G such that $\mathfrak{w} = e^h \mathfrak{v}$.

Proof. Take h such that $h' = g$. Then

$$(\mathfrak{w}\, e^{-h})' = e^{-h}(\mathfrak{w}' - h'\mathfrak{w}) \equiv 0$$

on G. Hence $\mathfrak{w} e^{-h} = \mathfrak{v}$ is constant on G,

q.e.d.

Lemma 7.6. Let $G \neq \emptyset$ be an open, connected subset of $\mathbb{C}$. Let V be a complex vector space of dimension $n + 1$. Let

$\mathfrak{w}: G \to V$ be a holomorphic vector function. Then $\mathfrak{w} \wedge \mathfrak{w}' \wedge \dots \wedge \mathfrak{w}^{(n)} \equiv 0$ if and only if there exists a linear subspace W of V with $W \neq V$ such that $\mathfrak{w}(G) \subseteq W$. (See Weyl [46], Wu [49]).

Proof. 1. If $W \neq V$ with $\mathfrak{w}(G) \subseteq W$ exists, let $\mathfrak{v}_0, \dots, \mathfrak{v}_n$ be a base of V such that $\mathfrak{v}_0, \dots, \mathfrak{v}_k$ is a base of W. Then $k \leqq n$. Also $\mathfrak{w} = w_0 \mathfrak{v}_0 + \dots + w_n \mathfrak{v}_n$ with $w_{k+1} = \dots = w_n = 0$. Obviously

$$\mathfrak{w} \wedge \mathfrak{w}' \wedge \dots \wedge \mathfrak{w}^{(n)} = (\det w_\mu^{(\nu)})\, \mathfrak{v}_0 \wedge \dots \wedge \mathfrak{v}_n \equiv 0.$$

2. Now, assume that $\mathfrak{w} \wedge \mathfrak{w}' \wedge \dots \wedge \mathfrak{w}^{(n)} \equiv 0$. If $\mathfrak{w} \equiv 0$, then $W = \{0\}$. Hence $\mathfrak{w} \not\equiv 0$ can be assumed. An integer k with $0 < k < n$ exists such that $\mathfrak{w}_p = \mathfrak{w} \wedge \dots \wedge \mathfrak{w}^{(k)} \not\equiv 0$ for $p = 0,1,\dots,k$ and $\mathfrak{w}_p \equiv 0$ for $p = k+1,\dots,n$. An open convex subset $G_0 \neq \emptyset$ of G exists such that $\mathfrak{w}_k(z) \neq 0$ for all $z \in G$. Then $\mathfrak{w}_k \wedge \mathfrak{w}^{(k+1)} = \mathfrak{w}_{k+1} \equiv 0$. Hence holomorphic functions g_μ exist uniquely on G_0 such that

$$\mathfrak{w}^{(k+1)} = \sum_{\mu=0}^{k} g_\mu \mathfrak{w}^{(\mu)}$$

on G_0. Therefore

$$\mathfrak{w}_k' = \mathfrak{w}_{k-1} \wedge \mathfrak{w}^{(k+1)} = \mathfrak{w}_{k-1} \wedge g_k \mathfrak{w}^{(k)} = \mathfrak{w}_k g_k.$$

According to Lemma 7.5 a holomorphic function h on G_0 and a constant vector $\mathfrak{v} \in {}_{k+1}\Lambda V$ exist such that $\mathfrak{w}_k = e^{-h}\mathfrak{v}$.

Then $\mathfrak{v} = e^{-h}\mathfrak{w}_k \in \tilde{G}_k(V)$. Hence $W = \{\mathfrak{z} \mid \mathfrak{z} \wedge \mathfrak{v} = 0\}$ is a linear subspace of dimension k+1 of V with $k < n$. Hence $W \neq V$. If $z \in G_0$, then

$$\mathfrak{w}(z) \wedge \mathfrak{v} = \mathfrak{w}(z) \wedge \mathfrak{w}_k(z)\, e^{-h} = 0.$$

By analytic continuation $\mathfrak{w} \wedge \mathfrak{v} \equiv 0$ on G. Therefore $\mathfrak{w}(G) \subseteq W$, q.e.d.

Lemma 7.6 implies that in the case $m = 1$ a holomorphic map $f: M \to \mathbb{P}(V)$ is general (for $B \equiv 1$) if and only if f is generic.

Assume (A1) and (A3). Let $\mathfrak{w}: M \to V$ be a holomorphic map. Then the following auxiliary construction shall be carried out. If $\emptyset \neq S \subseteq M$, let L(S) be the linear hull of $\mathfrak{w}(S)$ in V. The following properties are easily obtained.

1) $\mathfrak{w}(S) \subseteq L(S)$.

2) If W is a linear subspace of V with $\mathfrak{w}(S) \subseteq W$ then $L(S) \subseteq W$.

3) If $\dim L(S) = p$, then there exists a finite subset S_0 of S such that $L(S) = L(S_0)$ and such that $\# S_0 = p$.

4) If $\emptyset \neq S_1 \subseteq S_2 \subseteq M$, then $L(S_1) \subseteq L(S_2)$.

5) If $U \neq \emptyset$ is open in M, then $L(U) = L(M)$.

Proof of 5). Obviously, $L(U) \subseteq L(M)$. Define $p = \dim L(U)$. Then $0 \leqq p \leqq n+1$. A linear surjective map $\alpha: V \to \mathbb{C}^{n+1-p}$ exists such that L(U) is the kernel of α. The map $\alpha \circ \mathfrak{w}: M \to \mathbb{C}^{n+1-p}$ is holomorphic with $\alpha \circ \mathfrak{w} | U \equiv 0$. Because M is connected, $\alpha \circ \mathfrak{w} \equiv 0$. Hence $\mathfrak{w}(M) \subseteq \ker \alpha = L(U)$. Therefore $L(M) \subseteq L(U)$; q.e.d.

A subset N of M is said to be a complex curve if and only if there exists an open, connected subset $G \neq \emptyset$ of $\mathbb{C}$ and an injective, smooth, holomorphic map $\varphi\colon G \to M$ with $\varphi(G) = N$. Observe that N may not be closed in M and that $\varphi\colon G \to N$ may not be a homeomorphism. However φ is an imersion of G into M.

Lemma 7.7. Take $a \in M$. Assume $L(M) \geqq p \geqq 0$. Then there exists a complex curve N in M with $a \in N$ such that $\dim L(N) \geqq p$.

Proof. If $m = 1$, take $N = M$. Hence, assume $m > 1$. For $p = 0$ the assertion is trivial. Assume it is correct for p. Then it shall be proven form $p + 1$. Assume $\dim L(M) \geqq p+1$. By assumption an open, connected neighborhood G of $0 \in \mathbb{C}$ and an injective, smooth, holomorphic map $\psi\colon G \to M$ exist such that $\psi(0) = a$ and $\dim L(\psi(G)) \geqq p$. If $\dim L(\psi(G)) \geqq p+1$, take $N = \psi(G)$. Hence $\dim L(\psi(G)) = p$ can be assumed.

Write $E = L(\psi(G))$. A radius $r_0 > 0$ exists such that $\mathbb{C}(r_0) \subset G$ and such that there exists $s > 0$ and a biholomorphic map

$$\alpha\colon \mathbb{C}(r_0) \times (\mathbb{C}(s))^{m-1} \to U$$

onto an open neighborhood U of a in M such that $\alpha(z,0) = \psi(z)$ if $z \in \mathbb{C}(r_0)$ and $0 \in \mathbb{C}^{m-1}$. W.l.o.g. $r_0 = 2$ can be assumed. Then $E = L(\psi(\mathbb{C}(1)))$. Hence a set $S = \{s_1,\dots,s_p\}$ exists such that $E = L(S)$ and $s_\mu = \psi(z_\mu)$ with $|z_\mu| < 1$ for $\mu = 1,\dots,p$. Here $z_\mu \neq z_\nu$ if $\mu \neq \nu$. A holomorphic function $g\colon \mathbb{C} \to \mathbb{C}$ is defined by

$$g(z) = z \frac{(z-z_1) \ldots (z - z_p)}{(1-z_1) \ldots (1 - z_p)}$$

Observe $g(0) = g(z_1) = \ldots = g(z_p) = 0$ and $g(1) = 1$. A constant $C > 1$ exists such that $|g(z)| < C$ for all $|z| \leqq 2$. A holomorphic map

$$\mathfrak{x} = (x_1, \ldots, x_m) : \mathbb{C}^m \to \mathbb{C}^m$$

is defined by $\mathfrak{x}(z, \mathfrak{w}) = (z, g(z)\mathfrak{w})$ if $z \in \mathbb{C}$ and $\mathfrak{w} \in \mathbb{C}^{m-1}$. Then

$$\frac{\partial(x_1, \ldots, x_m)}{\partial(z, w_2, \ldots, w_m)} = (g(z))^{m-1}$$

Hence

$$\mathfrak{x}(1, \mathfrak{w}) = (1, \mathfrak{w})$$

$$\frac{\partial(x_1, \ldots, x_m)}{\partial(z_1 w_2, \ldots, w_m)}(1, \mathfrak{w}) = 1$$

for $\mathfrak{w} \in \mathbb{C}^{m-1}$. Define $r = \frac{s}{C} < s$. If $\mathfrak{w} = (w_2, \ldots, w_m) \in \mathbb{C}(r)^{m-1}$ and $z \in \mathbb{C}(2)$ then $|w_\mu \, g(z)| < s$. Hence

$$\mathfrak{x} : \mathbb{C}(2) \times (\mathbb{C}(r))^{m-1} \to \mathbb{C}(2) \times (\mathbb{C}(s))^{m-1}$$

Take $\mathfrak{b} = (b_2, \ldots, b_m) \in (\mathbb{C}(r))^{m-1}$, then $(1, \mathfrak{b}) = \mathfrak{v} \in \mathbb{C}(2) \times \mathbb{C}(r))^{m-1}$ open connected neighborhoods U_1 in $\mathbb{C}(2) \times (\mathbb{C}(r))^{m-1}$ and U_2 in $\mathbb{C}(2) \times (\mathbb{C}(s))^{m-1}$ of $\mathfrak{b}$ exist such that $\mathfrak{x} : U_1 \to U_2$ is biholomorphic. Then $U_3 = \alpha(U_2)$ is an open, connected neighborhood of $\alpha(\mathfrak{b})$ in U. Then $L(U_3) = L(M) \not\subseteq E$. Therefore $s_{p+1} \in U_3$ with $\mathfrak{v}(s_{p+1}) \notin E$ exists. Then $s_{p+1} = \alpha(\mathfrak{x}(z_{p+1},\mathfrak{n}))$ where $(z_{p+1},\mathfrak{n}) \in U_1$. Define $\mathfrak{y} : \mathbb{C}(2) \to \mathbb{C}(2) \times \mathbb{C}(s)^{m-1}$ by $\mathfrak{y}(z) = \mathfrak{x}(z,\mathfrak{n}) = (z, g(z)\mathfrak{n})$. Then $\mathfrak{y}$ is an injective, smooth holomorphic map. Also

$$\varphi = \alpha \circ \mathfrak{y} : \mathbb{C}(2) \to M$$

is injective, smooth and holomorphic. Here $\varphi(0) = \alpha(0, g(0)\mathfrak{n}) = \alpha(0) = a$. Hence $N = \varphi(\mathbb{C}(2))$ is a complex curve in M with $a \in N$. If $1 \leqq \mu \leqq m$, then

$$\varphi(z_\mu) = \alpha(z_\mu, g(z_\mu)\mathfrak{n}) = \alpha(z_\mu, 0) = \psi(z_\mu) = s_\mu \in N$$

$$\varphi(z_{p+1}) = \alpha(\mathfrak{x}(z_{p+1},\mathfrak{n})) = s_{p+1} \in N$$

Define $S_1 = S \cup \{s_{p+1}\} = \{s_1,\ldots,s_{p+1}\} \subseteq N$. Then $W = L(S_1) \supseteq L(S) = E$. Here W has at most dimension p + 1 and $\mathfrak{v}(s_{p+1}) \in W - E$ with dim E = p. Therefore dim W = p+1. Consequently, p + 1 = dim W $\leqq$ dim L(N), q.e.d.

The case p = n+1 implies immediately.

<u>Theorem 7.8</u>. Assume (A1) and (A3). Let $\mathfrak{v} : M \to V$ be a holomorphic map such that $\mathfrak{v}(M)$ is not contained in any proper linear subspace of V. Take $a \in M$. Then there exists a complex curve N in M with $a \in N$ such that $\mathfrak{v}(N)$ is not contained in any proper linear subspace of V.

An immediate consequence of Lemma 7.6 and Theorem 7.8 is the following Corollary.

Corollary 7.9. Under the assumptions of Theorem 7.8, there exist an open, connected neighborhood G of $0 \in \mathbb{C}$ and an injective, smooth, holomorphic map $\varphi: G \to M$ with $\varphi(0) = a$ such that $\mathfrak{w} = \mathfrak{v} \circ \varphi$ and

$$\mathfrak{w} \wedge \mathfrak{w}' \wedge \cdots \wedge \mathfrak{w}^{(n)} \not\equiv 0 \quad \text{on } G.$$

After these preparations, the construction of the holomorphic form B can begin. At first, it will be constructed locally.

Lemma 7.10. Assume (A1). Take $a \in M$. Let G be an open, connected neighborhood of $0 \in \mathbb{C}$. Let $\varphi: G \to M$ be an injective, smooth, holomorphic map with $\varphi(0) = a$. Then there exist first, $\alpha \in \mathfrak{T}_M$ with $a \in U_\alpha$, second, an open connected neighborhood G_0 of 0 in G, such that the restriction $\varphi: G_0 \to U_\alpha$ is proper, and third, a holomorphic differential form B of bidegree $(m-1,0)$ on U_α such that

$$(D^p_{B,\alpha}\, h) \circ \varphi = (h \circ \varphi)^{(p)}$$

on G_0 for all $h \in \mathrm{Hol}\,(U_\alpha)$ and all $0 \leqq p \in \mathbb{Z}$. Here $(h \circ \varphi)^{(p)}$ is the pth derivative on G_0.

Proof. Because φ is a holomorphic immersion, a patch $\alpha \in \mathfrak{T}_M$ exists such that $a \in U_\alpha$ and such that $\alpha: U_\alpha \to G_0 \times H$ is biholomorphic, where G_0 is an open, connected neighborhood of 0 in G, where H is an open connected neighborhood of

$0 \in \mathbb{C}^{m-1}$ and where $\alpha(\varphi(z)) = (z,0)$ for all $z \in G_0$ if $0 \in \mathbb{C}^{m-1}$. Then $\varphi: G_0 \to U_\alpha$ is proper. Let $z_1,\ldots,z_m$ be the coordinate functions of $\mathbb{C}^m$. Write $\alpha = (\alpha_1,\ldots,\alpha_m)$. Then $\alpha^*(dz_\mu) = d\alpha_\mu$. Define

$$B = d\alpha_2 \wedge\ldots\wedge d\alpha_m$$

on U_α. If $h \in \mathrm{Hol}\,(U_\alpha)$, then

$$(D_{B,\alpha}\, h)\, d\alpha_1 \wedge\ldots\wedge d\alpha_m = dh \wedge B = h_{\alpha_1}\, d\alpha_1 \wedge\ldots\wedge d\alpha_m$$

Hence

$$(D_{B,\alpha}\, h) \circ \alpha^{-1} = \frac{\partial}{\partial z_1}\,(h \circ \alpha^{-1})$$

on G_0. By iteration

$$(D^p_{B,\alpha}\, h) \circ \alpha^{-1} = \frac{\partial^p}{\partial z_1^p}\,(h \circ \alpha^{-1})$$

Observe $\varphi(z) = \alpha^{-1}(z,0)$. Therefore

$$(D^p_{B,\alpha}\, h)(\varphi(z)) = \frac{\partial^p}{\partial z_1^p}\,(h \circ \alpha^{-1})(z,0) = (h \circ \varphi)^{(p)}$$

q.e.d.

<u>Theorem 7.11</u>. Assume (A1), (A3) and (A4). Suppose that the meromorphic map $f: M \to \mathbb{P}(V)$ is generic. Let W be a complex vector space of dimension k. Assume a holomorphic map $\mathfrak{w}: M \to W$ and an open, connected subset $U \neq \emptyset$ of M exist such that $\mathfrak{w}: U \to W$ is injective and smooth. Then

there exists a holomorphic differential form $\hat{B}$ of bidegree $(m-1,0)$ on W, whose coefficients are polynomials of atmost degree $n-1$ on W such that f is general for $B = \mathfrak{w}^*(\hat{B})$.

Remark 1. A Stein manifold M can be holomorphically embedded into $\mathbb{C}^{2m+1} = W$. Hence such a map $\mathfrak{w}$ exists on each Stein manifold. Does $\mathfrak{w}$ exist, if M is pseudoconvex in the sense of section 10?

Remark 2. The growth of B will have to be estimated. Therefore the fact that $\hat{B}$ can be taken with polynomial coefficients of degree $\leqq n-1$ is of great help. Especially, this is true for the application to the Bezout Theorem where $M \subseteq V = W$ and where $\mathfrak{w}: M \to V$ is the inclusion.

Proof. The case $m = 1$ is trivial. Assume $m > 1$. W.l.o.g. $W = \mathbb{C}^k$. Let $z_1, \ldots, z_k$ be the coordinate functions on $\mathbb{C}^k$. The set U can be taken so small that there exists an irreducible representation $\mathfrak{v}: U \to V$ of f on M. Then $\mathfrak{v}(U)$ is not contained in any proper linear subspace of V because f is generic. According to Corollary 7.9 an open connected subset $G \neq \emptyset$ of $\mathbb{C}$ and an injective, smooth holomorphic map $\varphi: G \to M$ with $\mathfrak{y} = \mathfrak{v} \circ \varphi$ exists such that

$$\mathfrak{y} \wedge \mathfrak{y}' \wedge \cdots \wedge \mathfrak{y}^{(n)} \not\equiv 0 \text{ on } G.$$

W.l.o.g. $0 \in G$ and $\mathfrak{y} \wedge \mathfrak{y}' \wedge \cdots \wedge \mathfrak{y}^{(n)}(0) \neq 0$ can be assumed. Then $a = \varphi(0) \in U$. By Lemma 7.10, there exist first, $\alpha \in \mathfrak{S}_M$ with $a \in U_\alpha \subseteq U$, second, an open, connected neighborhood G_0 of $0 \in G$ such that $\varphi: G_0 \to U_\alpha$ is proper, and third, a holomorphic differential form B_0 of bidegree $(m-1,0)$ on U_α such that

$$(D^p_{B_0,\alpha}\, h) \circ \varphi = (h \circ \varphi)^{(p)}$$

on G_0 if $0 \leqq p \in \mathbb{Z}$ and $h \in \mathrm{Hol}\,(U_\alpha)$. Now

$$\mathfrak{w}_{B_0,n} = \mathfrak{w} \wedge D_{B_0,\alpha}\mathfrak{w} \wedge \cdots \wedge D^n_{B_0,n}\mathfrak{w}$$

$$\mathfrak{w}_{B_0,n}(a) = \mathfrak{y} \wedge \mathfrak{y}' \wedge \cdots \wedge \mathfrak{y}^{(n)}(0) \neq 0.$$

Because $\mathfrak{w} : U \to W$ is an immersion with $a \in U$, there exist open, connected neighborhood U_1 of a in U_α and H of $\mathfrak{v} = \mathfrak{w}(a)$ in W and a holomorphic map $\rho\colon H \to U_1$ such that $\mathfrak{w} : U_1 \to H$ is proper and such that $\rho \circ \mathfrak{w} = \mathrm{Id}\colon U_1 \to U_1$ is the identity map. Then $B^* = \rho^*(B_0)$ is a holomorphic differential form of bidegree $(m-1,0)$ on H with

$$\mathfrak{w}^*(B^*) = \mathfrak{w}^*(\rho^*(B_0)) = (\rho \circ \mathfrak{w})^*(B_0) = B_0$$

on U_1.

Let $\mathfrak{J}$ be the set of all increasing, injective maps

$$\nu\colon \mathbb{N}[1,m-1] \to \mathbb{N}[1,k]$$

For $\nu \in \mathfrak{J}$ define

$$dz_\nu = dz_{\nu(1)} \wedge \cdots \wedge dz_{\nu(m-1)}.$$

For $\mu = 1,\ldots,m$ define

$$\eta_\mu = (-1)^{\mu-1}\, d\alpha_1 \wedge\cdots\wedge d\alpha_{\mu-1} \wedge d\alpha_{\mu+1} \wedge\cdots\wedge d\alpha_m$$

Observe that $\mathfrak{w} = (w_1,\ldots,w_m) : M \to \mathbb{C}^k$. Hence holomorphic functions ζ_μ^ν exist such that for each $\nu \in \mathfrak{z}$

$$dw_\nu = \sum_{\mu=1}^{m} \zeta_\mu^\nu\, \eta_\mu$$

Let $\mathfrak{L}$ be the set of all holomorphic differential forms of bidegree $(m-1,0)$ on H. Take any $C^* \in \mathfrak{L}$ and define $C = \mathfrak{w}^*(C^*)$ on U_1. Then holomorphic functions C_ν^* exist on H such that

$$C^* = \sum_{\nu\in\mathfrak{z}} C_\nu^*\, dz_\nu$$

$$C = \sum_{\mu=1}^{m} \Big[\sum_{\nu\in\mathfrak{z}} (C_\nu^* \circ \mathfrak{w})\, \zeta_\mu^\nu\Big]\, \eta_\mu$$

If $h \in \mathrm{Hol}\,(U_1)$ then

$$D_{C,\alpha}\, h = \sum_{\mu=1}^{m} \sum_{\nu\in\mathfrak{z}} (C_\nu^* \circ \mathfrak{w})\, \zeta_\mu^\nu\, h_{\alpha\mu}$$

on U_1. Observe that the coefficients ζ_μ^ν do not depend on the choice of the form $C^* \in \mathfrak{L}$. A similar formula holds for the iterated operator. Define

$$A_p = \{(\mu_1,\ldots,\mu_m) \mid 0 \leqq \mu_\lambda \in \mathbb{Z},\ p \geqq \mu_1 +\ldots+ \mu_m\}$$

$$h_\alpha = h_{\alpha_1^{\mu_1} \cdots \alpha_m^{\mu_m}} \qquad \text{if } \mu \in A_p.$$

Then there exist holomorphic functions $\xi^p_{\rho\mu}$ on U_1 independent of $C^* \in \mathfrak{C}$ and polynomials $P^p_{\mu\rho}$ in the coefficients C^*_ν of $C \in \mathfrak{C}$ and their partial derivatives up and including order p-1 such that

$$D^p_{C,\alpha}\, h = \sum_{\mu \in A_p} \sum_{\rho=1}^{s_p} (P^p_{\mu\rho} \circ \mathfrak{w})\, \xi^p_{\rho\mu}\, h_{\alpha_\mu}$$

on U_1. This is easily seen by induction for p.

Take $r > 0$ such that $H_0 = \{\mathfrak{z} \in W \mid |\mathfrak{z} - \mathfrak{w}| < r\} \subseteq H$. Then

$$B^*_\nu(\mathfrak{z}) = \sum_{\mu=0}^{\infty} B^*_{\nu\mu}(\mathfrak{z} - \mathfrak{w})$$

converges uniformly on each compact subset of H_0. Here $B^*_{\nu\mu}$ is a homogeneous polynomial of degree μ, whose coefficients are the partial derivatives of order μ at $\mathfrak{w}$ of the function B^*_ν. Define

$$\hat{B}_\nu(\mathfrak{z}) = \sum_{\mu=0}^{n-1} B^*_{\nu\mu}\,(\mathfrak{z} - \mathfrak{w})$$

on $\mathbb{C}^k$. Then $\hat{B}_\nu$ is a polynomial of atmost degree n-1. Here $\hat{B}_\nu$ and B^*_ν share the same partial derivatives at $\mathfrak{w}$ for the orders 0,1,...,n-1. A holomorphic differential form

$$\hat{B} = \sum_{\nu \in \mathfrak{I}} \hat{B}_\nu\, dz_\nu$$

of bidegree (m-1,0) is defined on W. Define $B = \mathfrak{w}^*(\hat{B})$ on M. Then $(D^p_{B,\alpha} h)(a) = D^p_{B_0,\alpha}(h)(a)$ for all $p = 0,1,\ldots,n$ and for all $h \in \mathrm{Hol}(U_1)$. Therefore $\mathfrak{w}_{B,n}(a) = \mathfrak{w}_{B_0,n}(a) \neq 0$. The meromorphic map f is general for B, q.e.d.

VIII. THE PLÜCKER FORMULAS

The following, additional general assumptions shall be made

(A5) Let B be a holomorphic differential form of bidegree (m-1,0) on M. Let $h'_\alpha = D_{B,\alpha} h$ be the associated contravariant differential operator. (If m = 1, take $B \equiv 1$).

(A6) Assume that f is general for B.

The assumption (A6) implies that f is generic, hence $\mathfrak{A}_f = \mathbb{P}(V^*)$. Now, assume (A1) - (A6). Take a patch $\alpha \in \mathfrak{T}_M$ and assume a representation $\mathfrak{w} : U_\alpha \to V$ of f is defined. Then

$$\mathfrak{w}_{p,\alpha} = \mathfrak{w} \wedge \mathfrak{w}'_\alpha \wedge \cdots \wedge \mathfrak{w}^{(p)}_\alpha : U_\alpha \to \textstyle\bigwedge_{p+1} V$$

is a representation of f_p. Then the greatest common divisor $\delta_{\mathfrak{w}_{p,\alpha}}$ of $\mathfrak{w}_{p,\alpha}$ is defined on U_α. Assume that $\mathfrak{w}$ is irreducible. Let $\tilde{\mathfrak{w}} : U_\beta \to V$ be another irreducible representation of f on U_β where $\beta \in \mathfrak{T}_M$. Assume $U_\beta \cap U_\alpha \neq \emptyset$. Then $\tilde{\mathfrak{w}} = h\mathfrak{w}$ where $\tilde{h}: U_\alpha \cap U_\beta \to \mathbb{C} - \{0\}$ is holomorphic. Hence

$$\tilde{\mathfrak{w}}_{p\beta} = h^{p+1} (\Delta_{\alpha\beta})^{\frac{p(p+1)}{2}} \mathfrak{w}_{p\alpha}$$

$$\delta \tilde{\mathfrak{w}}_{p\beta} = \delta \mathfrak{w}_{p\alpha}$$

on $U_\alpha \cap U_\beta$. Therefore one and only one non-negative divisor $d_p \geqq 0$ exists on M such that $d_p = \delta \mathfrak{w}_{p\alpha}$ on U_α if $\alpha \in \mathfrak{T}_M$ and if $\mathfrak{w} : U_\alpha \to M$ is an irreducible representation of f on U_α. The divisor d_p is called the pth <u>greatest common divisor</u> of f. Obviously $d_0 \equiv 0$ and $d_p \leqq d_{p+1}$ for $p = 0,1,\ldots,n-1$. The pth <u>stationary divisor</u> v_p is defined by

$$v_p = d_{p+1} - 2d_p + d_{p-1} \quad \text{if } 1 \leqq p \leqq n-1$$

$$v_0 = d_1$$

<u>Lemma 8.1.</u>[4] Assume (A1) - (A6). Then $v_p \geqq 0$. If $\alpha \in \mathfrak{T}_M$ and if $\mathfrak{w} : U_\alpha \to V$ is a representation, then

$$v_p = \delta \mathfrak{w}_{p+1} - 2\delta \mathfrak{w}_p + \delta \mathfrak{w}_{p-1} \geqq 0 \quad \text{if } 1 \leqq p \leqq n-1$$

$$v_0 = \delta \mathfrak{w}_1 - 2\delta \mathfrak{w} .$$

Proof. Take $x_0 \in U_\alpha$. An open neighborhood U of x exists such that there are irreducible representations $\mathfrak{w}_q$ of f_q on U for $q = p-1, p, p+1$ and such that there is an irreducible representation $\mathfrak{y}$ of $(f_p)_1$. They do not depend on the choice

---

4) See [36] Satz 13.2.

of $\alpha \in \mathfrak{T}_M$ with $x \in U_\alpha$ and of $\mathfrak{w}: U_\alpha \to V$. Let $\wedge$ be the exterior product on the vector space $\underset{p+1}{\Lambda} V$. Then holomorphic functions D_q on $U \cap U_\alpha$ and E on U exist such that $\mathfrak{w}_q = D_q \mathfrak{u}_q$ and $\mathfrak{u}_p \wedge \mathfrak{u}'_{p\alpha} = E\,\mathfrak{y}$ on U. Here E does not depend on α and $\mathfrak{w}$. Observe $D_q = \delta_{\mathfrak{w}_{q,\alpha}}$. By (2.1)

$$|D_{p-1}|\ |D_{p+1}|\ |\mathfrak{u}_{p-1}|\ |\mathfrak{u}_{p+1}| = |\mathfrak{w}_{p-1,\alpha}|\ |\mathfrak{w}_{p+1,\alpha}|$$

$$= |\mathfrak{w}_{p\alpha} \wedge \mathfrak{w}'_{p\alpha}| = |D_p \mathfrak{u}_p \wedge (D_p \mathfrak{u}_p)'_\alpha| = |D_p|^2\ |\mathfrak{u}_p \wedge \mathfrak{u}'_{p\alpha}|$$

$$= |D_p|^2\ |E|\ |\mathfrak{y}|$$

Define the meromorphic function

$$F = \frac{D_{p-1}\ D_{p+1}}{E\ \ D_p^2}$$

on $U \cap U_\alpha$. Define $S = \mathfrak{y}^{-1}(0) \cup (\mathfrak{u}_{p-1})^{-1}(0) \cup (\mathfrak{u}_p)^{-1}(0) \cap U \cap U_\alpha$. Then

$$|F| = \frac{|\mathfrak{y}|}{|\mathfrak{u}_{p-1}|\ |\mathfrak{u}_{p+1}|} > 0 \quad \text{on } U \cap U_\alpha - S$$

Because $\dim S \leqq m-2$, the function F is holomorphic on $U \cap U_\alpha$ with $F(z) \neq 0$ for all $z \in U \cap U_\alpha$. Therefore

$$0 \equiv \delta_F = \delta_{D_{p+1}} - 2\delta_{D_p} + \delta_{D_{p-1}} - \delta_E$$

or

$$0 \leqq \delta_E = \delta_{\mathfrak{w}_{p+1,\alpha}} - 2\delta_{\mathfrak{w}_{p,\alpha}} + \delta_{\mathfrak{w}_{p-1,\alpha}}$$

on $U \cap U_\alpha$. Here E is independent of the choice of α and $\mathfrak{w}$; hence $\mathfrak{w}$ can be taken irreducible. Therefore

$$0 \leqq \delta_E = d_{p+1} - 2d_p + d_{p-1} = v_p$$

on $U \cap U_\alpha$;

q.e.d.

The map $f_p\colon M \to \mathbb{P}(V)$ is meromorphic and holomorphic on exactly $M - I_{f_p}$. Abbreviate

$$i_0 = \left(\frac{i}{2\pi}\right)^{m-1} (-1)^{\frac{(m-1)(m-2)}{2}} = \left(\frac{1}{2\pi}\right)^{m-2} i^{((m-1)^2)} \quad (8.1)$$

A non-negative, real analytic differential form

$$H_p = i_0\, B \wedge \overline{B} \wedge f_p^*(\ddot{\omega}_p) \geqq 0$$

is defined on $M - I_{f_p}$.

Lemma 8.1. Assume (A1) - (A6). Let $\alpha = (\alpha_1, \ldots, \alpha_m) \in \mathfrak{T}_M$ be a patch of M and let $\mathfrak{w}: U_\alpha \to V$ be a representation of f. If $0 < p < n$, define

$$H_p^\alpha = |\mathfrak{w}_{p-1,\alpha}|^2\, |\mathfrak{w}_{p+1,\alpha}|^2\, |\mathfrak{w}_{p,\alpha}|^{-4}$$

$$H_0^\alpha = |\mathfrak{w}_{1,\alpha}|^2\, |\mathfrak{w}|^{-4}$$

on $U_\alpha - \mathfrak{w}_{p\alpha}^{-1}(0)$ (respectively on $U_\alpha - \mathfrak{w}^{-1}(0)$). Then

$$H_p = H_p^\alpha \left(\frac{i}{2\pi}\right)^m d\alpha_1 \wedge d\bar{\alpha}_1 \wedge \dots \wedge d\alpha_m \wedge d\bar{\alpha}_m$$

on $U_\alpha - \mathfrak{w}_{p\alpha}^{-1}(0)$ for $p = 0,1,\dots,n-1$.

Proof. Define $\mathfrak{w}_{-1,\alpha} \equiv 1$. Define

$$\upsilon_\alpha = \left(\frac{i}{2\pi}\right)^m d\alpha_1 \wedge d\bar{\alpha}_1 \wedge \dots \wedge d\alpha_m \wedge d\bar{\alpha}_m \tag{8.2}$$

Then

$$i_0\, B \wedge \bar{B} \wedge f_p^*(\ddot{\omega}_p) = \frac{i}{2\pi}\, i_0\, B \wedge \bar{B} \wedge \partial\bar{\partial} \log |\mathfrak{w}_p|^2$$

$$= |\mathfrak{w}_p|^{-4} \sum_{\mu|\nu=1}^{m} [\,|\mathfrak{w}_p|^2 (\mathfrak{w}_{p\alpha_\mu} | \mathfrak{w}_{p\alpha_\nu}) - (\mathfrak{w}_{p\alpha_\mu} | \mathfrak{w}_p)(\mathfrak{w}_p | \mathfrak{w}_{p\alpha_\nu})\,] b_\mu \bar{b}_\nu \upsilon_\alpha$$

$$= |\mathfrak{w}_p|^{-4}\, [\,|\mathfrak{w}_p|^2 |\mathfrak{w}_p'|^2 - |(\mathfrak{w}_p' \,|\, \mathfrak{w}_p)|^2\,]\, \upsilon_\alpha$$

$$= |\mathfrak{w}_p|^{-4}\, |\mathfrak{w}_p \wedge \mathfrak{w}_p'|^2\, \upsilon_\alpha$$

$$= |\mathfrak{w}_p|^{-4}\, |\mathfrak{w}_{p-1}|^2\, |\mathfrak{w}_{p+1}|^2\, \upsilon_\alpha$$

$$= H_p^\alpha\, \upsilon_\alpha$$

q.e.d.

Historically, this identity prompted the introduction of B.

Let K be the canonical bundle of M. Let κ be a hermitian metric along the fibers of K. If $\alpha = (\alpha_1,\dots,\alpha_m) \in \mathfrak{T}_M$ is a patch of M, then $d\alpha_1 \wedge \dots \wedge d\alpha_m$ is a holomorphic frame of K over U_α. Hence

$$\kappa_\alpha = \kappa(d\alpha_1 \wedge \dots \wedge d\alpha_m)$$

is a positive function of class C^∞ on U_α. If $\beta \in \mathfrak{V}_M$ with $U_\alpha \cap U_\beta \neq \emptyset$ then

$$d\alpha_1 \wedge \cdots \wedge d\alpha_m = \Delta_{\alpha\beta}\, d\beta_1 \wedge \cdots \wedge d\beta_m$$

$$\kappa_\alpha = |\Delta_{\alpha\beta}|^2\, \kappa_\beta$$

on $U_\alpha \cap U_\beta$. Define the Ricci form of κ by

$$\text{Ric}\ \kappa = -c_1(K,\kappa)$$

on U_α

$$\text{Ric}\ \kappa = \frac{1}{4\pi}\, dd^\perp \log \kappa_\alpha = -c_1(K,\kappa).$$

On $U_\alpha \cap U_\beta$, the identity $H_p^\alpha = |\Delta_{\beta\alpha}|^2\, H_p^\beta$ holds. Hence $H_p^\alpha\, \kappa_\alpha = H_p^\beta \kappa_\beta$ on $U_\alpha \cap U_\beta$. Therefore a function $\kappa\, H_p$ of class C^∞ exists uniquely on $M - I_{f_p}$ such that for each $\alpha \in \mathfrak{V}_M$ and each representation $\mathfrak{w} : U_\alpha \to V$, the identity $\kappa\, H_p = H_p^\alpha\, \kappa_\alpha$ holds on $U_\alpha - (\mathfrak{w}_{p,\alpha})^{-1}(0)$.

Let (G,g,ψ) be a bump on M. Define

$$N_{v_p}(G) = \int_{G \cap \gamma(v_p)} v_p\, \psi\, \chi \geqq 0$$

$$\text{Ric}_\kappa(G) = \int_G \psi\, (\text{Ric}\ \kappa) \wedge \chi$$

$$\Omega_f^p(dG,\kappa) = \frac{1}{4\pi} \int_{dG} \log\, (\kappa\, H_p)\, d^\perp\psi \wedge \chi$$

$$\Omega_f^p(dg,\kappa) = \frac{1}{4\pi} \int_{dg} \log (\kappa\, H_p)\, d^{\perp}\psi \wedge \chi$$

$$S_f^p(G,\kappa) = \frac{1}{4\pi} \int_{G-g} \log \frac{1}{\kappa\, H_p}\, dd^{\perp}\psi \wedge \chi$$

Obviously, the two first integrals exist. The existence of the other is a consequence of Theorem 8.3 below. $N_{V_p}(G)$ is called the pth stationary valence of f and $\mathrm{Ric}_{\kappa}(G)$ is called the Ricci function. It is independent of f and B.

Theorem 8.3. Plücker's difference formula. Assume (A1) - (A6). Let κ be a hermitian metric along the fibers of the canonical bundle of M. Let (G,g,ψ) be a bump on M. Take $0 \leqq p \leqq n-1$ and define $T_{f_{-1}} = 0$. Then

$$V_p(G) + T_{f_{p+1}}(G) - 2T_{f_p}(G) + T_{f_{p-1}}(G) =$$

$$= \Omega_f^p(dG,\kappa) - \Omega_f^p(dg,\kappa) + S_f^p(G,\kappa) + \mathrm{Ric}_{\kappa}(G)$$

Remark 1. Weyl [48] and Wu [51] called this formula the second main theorem. It corresponds to the theorem on the logarithmic derivative in the classical case.

Remark 2. The left side does not depend on κ and each term in the expression is non-negative. On the right hand side each term depends on κ and nothing can be said about the signs of the terms.

Proof. Only the case $p > 0$ shall be considered. The case $p = 0$ is proven exactly the same way with trivial notational changes. Select a family $\{\alpha_\lambda\}_{\lambda\in\Lambda}$ of patches

$\alpha_\lambda \in \mathfrak{T}_M$ with a family $\{\mathfrak{w}_\lambda\}_{\lambda \in \Lambda}$ of irreducible representations $\mathfrak{w}_\lambda: U_\lambda \to V$ of f such that $U'_{\alpha_\lambda} = \mathbb{C}^m(1)$ is the unit ball for each $\lambda \in \Lambda$ and such that $M = \bigcup_{\lambda \in \Lambda} U_{\alpha_\lambda}$. Abbreviate $U_\lambda = U_{\alpha_\lambda}$ and $U_{\lambda_0 \ldots \lambda_p} = U_{\lambda_0} \cap \ldots \cap U_{\lambda_p}$ and $\mathfrak{w}_{q\lambda} = \mathfrak{w}_{q\alpha_\lambda}$. Define

$$\Lambda[p] = \{(\lambda_0, \ldots, \lambda_p) \mid U_{\lambda_0 \ldots \lambda_p} \neq \emptyset\}$$

If $(\lambda, \mu) \in \Lambda[1]$, then

$$d\alpha_{\lambda 1} \wedge \ldots \wedge d\alpha_{\lambda m} = \Delta_{\lambda\mu}\, d\alpha_{\mu 1} \wedge \ldots \wedge d\alpha_{\mu m}$$

on $U_{\lambda\mu}$. A holomorphic function $h_{\lambda\mu}: U_{\lambda\mu} \to \mathbb{C} - \{0\}$ exists such that $\mathfrak{w}_\lambda = h_{\lambda\mu} \mathfrak{w}_\mu$ on $U_{\lambda\mu}$. Then

$$\mathfrak{w}_{q\lambda} = h_{\lambda\mu}^{q+1} (\Delta_{\mu\lambda})^{\frac{q(q+1)}{2}} \mathfrak{w}_{\mu q} \quad \text{on } U_{\lambda\mu}$$

According to Lemma 4.1, $b_q \in \mathbb{P}(\textstyle\bigwedge_{q+1} V)$ exists such that f_q is adapted to b_q for $q = 0, 1, \ldots, n$. Take $\beta_q \in \mathbb{P}^{-1}(b_q)$ with $|\beta_q| = 1$. On U_λ, a meromorphic function r_λ is defined by

$$r_\lambda = (\beta_{p-1} \circ \mathfrak{w}_{p-1,\lambda})(\beta_{p+1} \circ \mathfrak{w}_{p+1,\lambda})\ (\beta_p \circ \mathfrak{w}_{p\lambda})^{-2}$$

Observe that the divisor of $\beta_q \circ \mathfrak{w}_{q,\lambda}$ is $(d_q + \delta_{f_q}^{b_q})|U_\lambda$. Abbreviate $\delta_{f_q}^{b_q} = \delta_q$. Hence

$$\delta_{r_\lambda} = d_{p-1} + \delta_{p-1} + d_{p+1} + \delta_{p+1} - 2d_p - 2\delta_p$$

$$= v_p + \delta_{p+1} - 2\delta_p + \delta_{p+1}$$

on U_λ. Therefore $\delta_r = v_p + \delta_{p+1} - 2\delta_p + \delta_{p+1}$ is a divisor on M such that $\delta_r | U_\lambda = \delta_{r_\lambda}$ for each $\lambda \in \Lambda$.

If $(\lambda,\mu) \in \Lambda[1]$, then $r_\lambda = r_\mu \Delta_{\mu\lambda}$ on $U_{\lambda\mu}$. Hence $\{r_\lambda\}_{\lambda\in\Lambda}$ could be considered a meromorphic section of the canonical bundle K. Since the Gauss-Bonnet formula was proved only for holomorphic sections, a slight detour will be necessary.

Because $\mathbb{C}^m(1) = U'_{\alpha_\lambda}$, holomorphic functions s_λ and t_λ exist on U_λ which are coprime at every point of U_λ, such that $t_\lambda r_\lambda = s_\lambda$. If $(\lambda,\mu) \in \Lambda[1]$, one and only one meromorphic function $g_{\lambda\mu}$ exists on $U_{\lambda\mu}$ such that

$$t_\lambda = g_{\lambda\mu}\, t_\mu$$

$$s_\lambda = g_{\lambda\mu}\, s_\mu\, \Delta_{\mu\lambda}$$

on $U_{\lambda\mu}$. The set $S_\lambda = \{x \in U_\lambda | s_\lambda(x) = t_\lambda(x) = 0\}$ is analytic with $\dim S_\lambda \leqq m-2$. The function $g_{\lambda\mu}$ is holomorphic and nowhere zero on $U_{\lambda\mu} - (S_\lambda \cup S_\mu)$. Hence $g_{\lambda\mu}\colon U_{\lambda\mu} \to \mathbb{C} - \{0\}$ is holomorphic. Obviously $g_{\lambda\mu} \cdot g_{\mu\lambda} = 1$ on $U_{\lambda\mu}$ and $g_{\lambda\mu} g_{\mu\rho} g_{\rho\lambda} = 1$ on $U_{\lambda\mu\rho}$ if $(\lambda,\mu,\rho) \in \Lambda[2]$. A holomorphic line bundle E on M exists which has a holomorphic frame e_λ on U_λ for each $\lambda \in \Lambda$ such that $e_\lambda = g_{\mu\lambda} e_\mu$ on $U_{\lambda\mu}$ for each $(\lambda,\mu) \in \Lambda[1]$. Hence $t_\lambda e_\lambda = t_\mu e_\mu$ on $U_{\lambda\mu}$. One and only one holomorphic section t of E exists such that $t | U_\lambda = t_\lambda e_\lambda$. Abbreviate $\upsilon_\lambda = d\alpha_{1\lambda} \wedge \cdots \wedge d\alpha_{m\lambda}$. Then $e_\lambda \otimes \upsilon_\lambda$ is a holomorphic frame of $E \otimes K$ over U_λ. Also $s_\lambda e_\lambda \otimes \upsilon_\lambda = s_\mu e_\mu \otimes \upsilon_\mu$ on $U_{\lambda\mu}$. One and only one holomorphic section s of $E \otimes K$ on M exists such that $s | U_\lambda = s_\lambda e_\lambda \otimes \upsilon_\lambda$ for each $\lambda \in \Lambda$. Then

$\delta_s | U_\lambda = \delta^0_{r_\lambda}$ and $\delta_t | U_\lambda = \delta^\infty_{r_\lambda}$. Hence

$$\delta_s - \delta_t = \delta_r = v_p + \delta_{p+1} - 2\delta_p + \delta_{p+1}$$

Abbreviate $N_{f_q}(G,b_q) = N_{\delta_q}(G) = N_q$ and $m_{f_q}(dG,b_q) = m_q$ and $m^0_{f_q}(dg,b_q) = m^0_q$ and $D_{f_q}(G,b_q) = D_q$. Then

$$N_{\delta_s}(G) - N_{\delta_t}(G) = N_{v_p}(G) + N_{p-1} - 2N_p + N_{p+1}$$

Let h be a hermitian metric along the fibers of E. Then $h(e_\lambda) = h_\lambda > 0$ is a function of class C^∞ on U_λ. One and only one hermitian metric $h\kappa$ along the fibers of $E \otimes K$ exists such that $h\kappa(e_\lambda \otimes v_\lambda) = h_\lambda \cdot \kappa_\lambda > 0$. Theorem 5.3 implies

$$N_{\delta_s}(G) = \frac{1}{4\pi} \int_{dG} \log h\kappa(s)\, d^\perp\psi \wedge \chi - \frac{1}{4\pi} \int_{dg} \log h\kappa(x)\, d^\perp\psi \wedge \chi$$

$$- \frac{1}{4\pi} \int_{G-g} \log h(s)\, dd^\perp\psi \wedge \chi - \int_G \psi\, c_1(E,h\kappa) \wedge \chi$$

$$N_{\delta_t}(G) = \frac{1}{4\pi} \int_{dG} \log h(t)\, d^\perp\psi \wedge \chi - \frac{1}{4\pi} \int_{dg} \log h(t)\, d^\perp\psi \wedge \chi$$

$$- \frac{1}{4\pi} \int_{G-g} \log h(t)\, dd^\perp\psi \wedge \chi - \int_G \psi\, c_1(E,h) \wedge \chi$$

If $\lambda \in \Lambda$, then

$$c_1(E,h\kappa) - c_1(E,h) = \frac{1}{4\pi}\, dd^c \log h_\lambda\kappa_\lambda - \frac{1}{4\pi}\, dd^c \log h_\lambda$$

$$= \frac{1}{4\pi}\, dd^c \log \kappa_\lambda = - \operatorname{Ric} \kappa$$

on U_λ. Also

$$\frac{h\kappa(s)}{h(t)} = \frac{h_\lambda \kappa_\lambda |s_\lambda|^2}{h_\lambda |t_\lambda|^2} = \kappa_\lambda \ |r_\lambda|^2$$

$$= \kappa_\lambda \ |\beta_{p-1} \circ \mathfrak{w}_{p-1,\lambda}|^2 \ |\beta_{p+1} \circ \mathfrak{w}_{p+1,\lambda}|^2 \ |\beta_p \circ \mathfrak{w}_{p,\lambda}|^{-4}$$

$$= ||b_{p-1};f_{p-1}||^2 \ ||b_{p+1};f_{p+1}||^2 \ ||b_p;f_p||^{-4} \ \kappa(H_p)$$

By subtraction

$$N_{v_p}(G) + N_{p-1} - 2N_p + N_{p+1} = N_{\delta_s}(G) - N_{\delta_t}(G) =$$

$$= \Omega_f^p \ (dG,\kappa) - \Omega_f^p \ (dg,\kappa) + S_f^p(G,\kappa) + \mathrm{Ric}_\kappa(G)$$

$$- m_{p-1} + 2m_p - m_{p-1} + m_{p-1}^0 - 2m_p^0 + m_{p+1}^0$$

$$+ D_{p-1} - 2D_p + D_{p+1}$$

Because

$$T_q(G) = N_q + m_q - m_q^0 - D_q$$

for $q = p-1,p,p+1$, the Plücker formula follows; q.e.d.

IX. THE AHLFORS ESTIMATES

If $m > 1$, a new general assumption will have to be made.

(A7) Assume that a bump (G,g,ψ) on M is given. Assume that $i_0 B \wedge \overline{B} \leqq \chi$ where i_0 is defined by (8.1).

If m = 1, then $B = \chi = 1$ and the assumption is satisfied. Recall that $\mathfrak{A}_f$ was the set of all $a \in \mathbb{P}(V^*)$ such that f is adapted to a.

Proposition 9.1. Assume (A1) - (A5) and (A7). Assume that f is general of order 1 for B. Then a constant $c_n > 2$ exists depending only on n and increasing with n such that for all λ with $0 < \lambda < 1$ and all $a \in \mathfrak{A}_f$ the following estimate holds.

$$\int_G \frac{||f_1;a||^2}{||f;a||^{2\lambda}} \psi H_0 \leqq \frac{c_n}{(1-\lambda)^2} [T_f(G) + \Phi^+(G)]$$

Proof. Take $a \in \mathfrak{A}_f$.

$$i_0 B \wedge \overline{B} \wedge f^*(\xi_a) \leqq f^*(\xi_a) \wedge \chi \quad \text{on } G.$$

Take $x_0 \in M$. Take $\beta \in \mathfrak{T}_M$ with $x_0 \in U_\beta$, such that an irreducible representation $\mathfrak{w}: U_\beta \to V$ of f on U_β exists. Take $\alpha \in \mathbb{P}^{-1}(a)$ with $|\alpha| = 1$. Define υ_α by (8.2). Then

$$i_0 B \wedge \overline{B} \wedge f^*(\xi_a) =$$

$$= i_0 B \wedge \overline{B} \left[\frac{i}{2\pi} |\mathfrak{w}|^2 \partial \frac{\alpha \circ \mathfrak{w}}{|\mathfrak{w}|^2} \wedge \overline{\partial} \frac{\overline{\alpha \circ \mathfrak{w}}}{|\mathfrak{w}|^2} + ||f;a||^2 f^*(\ddot{\omega}_0)\right]$$

$$= i_0 B \wedge \overline{B} \wedge \frac{i}{2\pi} |\mathfrak{w}|^{-4} [|\mathfrak{w}|^2 \partial(\alpha \circ \mathfrak{w}) \wedge \overline{\partial}\, \overline{\alpha \circ \mathfrak{w}} - (\alpha \circ \mathfrak{w})\, \partial |\mathfrak{w}|^2$$
$$\wedge \overline{\partial}\, \overline{\alpha \circ \mathfrak{w}} - \overline{\alpha \circ \mathfrak{w}}\, \partial(\alpha \circ \mathfrak{w}) \wedge \overline{\partial} |\mathfrak{w}|^2 + |\alpha \circ \mathfrak{w}|^2 \partial\overline{\partial} |\mathfrak{w}|^2]$$

$$= |\mathfrak{w}|^{-4} \sum_{\mu|\nu=1}^{m} [\, |\mathfrak{w}|^2 (\alpha \circ \mathfrak{w}_{\beta_\mu})(\alpha \circ \mathfrak{w}_{\beta_\nu}) - (\alpha \circ \mathfrak{w})(\mathfrak{w}_{\beta_\mu} | \mathfrak{w}) \overline{\alpha \circ \mathfrak{w}_{\beta_\nu}}$$

$$- \overline{\alpha \circ \mathfrak{w}} \; (\alpha \circ \mathfrak{w}_{\beta_\mu})(\mathfrak{w} | \mathfrak{w}_{\beta_\mu}) + |\alpha \circ \mathfrak{w}|^2 (\mathfrak{w}_{\beta_\mu} | \mathfrak{w}_{\beta_\nu})] b_\mu \overline{b}_\nu \, \upsilon_\alpha$$

$$= |\mathfrak{w}|^{-4} \; [\, |\mathfrak{w}|^2 |\alpha \circ \mathfrak{w}_\beta'|^2 - (\alpha \circ \mathfrak{w})(\mathfrak{w}_\beta' | \mathfrak{w})(\overline{\alpha \circ \mathfrak{w}_\beta'})$$

$$- \overline{\alpha \circ \mathfrak{w}} \; \overline{\alpha \circ \mathfrak{w}_\beta'} \; (\mathfrak{w} | \mathfrak{w}_\beta') + |\alpha \circ \mathfrak{w}|^2 \, |\mathfrak{w}_\beta'|^2] \; \upsilon_\alpha$$

$$= |\mathfrak{w}|^{-4} \; | \mathfrak{w}(\alpha \circ \mathfrak{w}_\beta') - \mathfrak{w}_\beta'(\alpha \circ \mathfrak{w})|^2 \; \upsilon_\alpha$$

$$= |\mathfrak{w}|^{-4} \; |(\mathfrak{w} \wedge \mathfrak{w}_\beta') \llcorner \alpha|^2 \; \upsilon_\alpha$$

$$= \frac{|(\mathfrak{w} \wedge \mathfrak{w}_\beta') \llcorner \alpha|^2}{|\mathfrak{w} \wedge \mathfrak{w}_\beta'|^2} \quad \frac{|\mathfrak{w} \wedge \mathfrak{w}_\beta'|^2}{|\mathfrak{w}|^4} \quad \upsilon_\alpha$$

$$= ||f_1;a||^2 \; H_0 \quad \text{on } U_\beta$$

Hence

$$i_0 \; B \wedge \overline{B} \wedge f^*(\xi_a) = ||f_1;a||^2 \; H_0$$

on $M - I_f$. Theorem 6.8 implies

$$\int_G \frac{||f_1;a||^2}{||f;a||^{2\lambda}} \, \psi \, H_0 = \int_G \frac{\psi}{||f;a||^{2\lambda}} \; i_0 \; B \wedge \overline{B} \wedge f^*(\xi_a)$$

$$\leq \int_G \frac{\psi}{||f;a||^{2\lambda}} \; f^*(\xi_0) \wedge \chi$$

$$\leq \frac{c_n}{(1-\lambda)^2} \; [T_f(G) + \Phi^+(G)]$$

q.e.d.

Assume (A1) - (A6). Let p and q be integers with $0 \leqq q < p \leqq n$. Take $b \in G_q(V^*)$. Take $\beta \in \mathbb{P}^{-1}(b)$. Then

$$E_p[b] = \{ \xi \in \tilde{G}_p(V) \mid \xi \llcorner \beta = 0 \}$$

is a thin analytic subset of $\tilde{G}_p(V)$ which is independent of the choice of β in $\mathbb{P}^{-1}(b)$. If $\xi \in E_p[b]$ and $0 \neq z \in \mathbb{C}$, then $z\xi \in E_p[b]$. Hence

$$\ddot{E}_p[b] = \mathbb{P}(E_p[b])$$

is a thin analytic subset of $G_p(V)$. If $x \in G_p(V)$, then $x \llcorner a$ is defined if and only if $x \in \ddot{E}_p[b]$. The map $x \to x \llcorner b$ is holomorphic.

The meromorphic map f_p is said to be <u>adjusted</u> to b if and only if $f_p^{-1}(\ddot{E}_p[b]) \neq M$. Obviously, f_p is adjusted to b, if and only if $\alpha \in \mathfrak{T}_M$ and $\beta \in \mathbb{P}^{-1}(b)$ and a representation $\mathfrak{w}: U_\alpha \to M$ exists such that $\mathfrak{w}_{p,\alpha} \llcorner \beta \not\equiv 0$ on U_α. If f_b is adjusted to b, then $\mathfrak{w}_{p,\alpha} \llcorner \beta \not\equiv 0$ on each open subset of U_α for all permissible choices of $\alpha, \beta, \mathfrak{w}$. Therefore, if f_p is adjusted to b, a meromorphic map

$$f_p \llcorner b \colon M \to \mathbb{P}({}_{p-q}\Lambda V)$$

is defined, which is holomorphic on $M - (I_{f_p} \cup f^{-1}(\ddot{E}_p[b]))$.

<u>Lemma 9.2</u>. Assume (A1) - (A6). Let p and q be integers with $0 \leqq q < p \leqq n$. Then the set D_{pq} of all $b \in G_q(V^*)$ such that f_p is not adjusted to b is thin in $G_q(V^*)$.

Proof. Take $x_0 \in M$. Take $\alpha \in \mathfrak{T}_M$ with $x_0 \in U_\alpha$ such that an irreducible representation $\mathfrak{w} : U_\alpha \to V$ of f on U_α exists. Then $x_1 \in U_\alpha$ exists such that $\mathfrak{y} = \mathfrak{w}_{p,\alpha}(x_1) \neq 0$. Hence

$$A = \{\beta \in \tilde{G}_q(V^*) \mid \mathfrak{y} \llcorner \beta = 0\}$$

is a thin analytic subset of $\tilde{G}_q(V^*)$. If $\beta \in A$ and $0 \neq z \in \mathbb{C}$, then $z\,\xi \in A$. Hence $C = \mathbb{P}(A)$ is a thin analytic subset of $G_q(V^*)$. If $b \in G_q(V^*) - C$, take $\beta \in \mathbb{P}^{-1}(b)$. Then $\mathfrak{y} \llcorner \beta \neq 0$ implies $\mathfrak{w}_{p,\alpha} \llcorner \beta \not\equiv 0$ on U_α. Hence f_p is adjusted to b. Therefore $b \in G_q(V^*) - D_{pq}$. Consequently $D_{pq} \subseteq C$ is thin; q.e.d.

Abbreviate $T_{f_p \llcorner b}(G) = T_p(G;b)$.

Theorem 9.3. Assume (A1) - (A7). Let p be an integer with $0 < p < n$. Take $b \in G_{p-1}(V^*)$ and $a \in \mathbb{P}(V^*)$ such that $b \wedge a$ exists. Assume, that f_p is adjusted to b and that $f_p \llcorner b$ is adjusted to a. then

$$\int_G \frac{||b:a||^2\, ||f_{p-1};b||^2\, ||f_{p+1};b \wedge a||^2}{||f_p \llcorner b;a||^{2\lambda}\, ||f_p;b||^4}\, H_p \leq \frac{c_n}{(1-\lambda)^2}\, \{T_p(G;b) + \Phi^+(G)\}$$

(Here, $||f_n;b \wedge a|| \equiv 1$).

Proof. Take $x_0 \in M$. Take $\alpha \in \mathbb{P}^{-1}(a)$ and $\beta \in \mathbb{P}^{-1}(b)$ and $\gamma \in \mathfrak{T}_M$, such that an irreducible representation $\mathfrak{w}: U_\gamma \to M$ of f over M exists. At first assume that $f_p \llcorner b$ is general of order 1 for b. Let H_0^* be the H_p formed for $f_p \llcorner b: M \to \mathbb{P}(V)$. Then 2.1 implies

$$|| (f_p | b)_1 ; a ||^2 \; H_0^*$$

$$= \frac{|\mathbf{\omega}_p \, \llcorner \, \beta \wedge (\mathbf{\omega}_p \, \llcorner \, \beta)' \, \llcorner \, \alpha|^2}{|(\mathbf{\omega}_p \, \llcorner \, \beta) \wedge (\mathbf{\omega}_p \, \llcorner \, \beta)'|^2 |\alpha|^2} \cdot \frac{|(\mathbf{\omega}_p \, \llcorner \, \beta) \wedge (\mathbf{\omega}_p \, \llcorner \, \beta)'|^2}{|\mathbf{\omega}_p \, \llcorner \, \beta|^4} \; \upsilon_\gamma$$

$$= |\alpha|^{-2} \; |\mathbf{\omega}_p \, \llcorner \, \beta|^{-4} \; |[(\mathbf{\omega}_p \, \llcorner \, \beta) \wedge (\mathbf{\omega}_{p-1} \wedge \mathbf{\omega}^{(p+1)} \, \llcorner \, \beta)] \, \llcorner \, \alpha|^2 \upsilon_\gamma$$

$$= |\alpha|^2 \; |\mathbf{\omega}_p \, \llcorner \, \beta|^{-4} \; |(\mathbf{\omega}_{p-1} ; \beta)(\mathbf{\omega}_{p+1} \, \llcorner \, \beta) \, \llcorner \, \alpha|^2 \; \upsilon_\gamma$$

$$= \frac{|(\mathbf{\omega}_{p-1} ; \beta)|^2}{|\mathbf{\omega}_{p-1}|^2 |\beta|^2} \cdot \frac{|\mathbf{\omega}_{p+1} \, \llcorner \, \beta \wedge \alpha|^2}{|\mathbf{\omega}_{p+1}|^2 |\beta \wedge \alpha|^2} \cdot \frac{|\mathbf{\omega}_p|^4 |\beta|^4}{|\mathbf{\omega}_p \, \llcorner \, \beta|^4} \cdot \frac{|\beta \wedge \alpha|^2}{|\beta|^2 |\alpha|^2} \cdot \frac{|\mathbf{\omega}_{p-1}|^2 |\mathbf{\omega}_{p+1}|^2}{|\mathbf{\omega}_p|^4} \upsilon_\gamma$$

$$= ||f_{p-1} ; b||^2 \; ||f_{p+1} ; b \wedge a||^2 \; ||f_p ; b||^{-4} \; ||b : a||^2 \; H_p$$

Proposition 9.1 implies

$$\int_G \psi \; \frac{||b : a||^2 \; ||f_{p-1} ; b||^2 \; ||f_{p+1} ; b||^2}{||f_p \, \llcorner \, b ; a||^{2\lambda} \; ||f_p ; b||^4} \; H_p$$

$$= \int_G \psi \; \frac{|| (f_p \, \llcorner \, b)_1 ; a ||^2}{||f_p \, \llcorner \, b ; a||^{2\lambda}} \; H_0^* \le \frac{c_n}{(1-\lambda)^2} \; [T_p(G ; b) + \Phi^+(G)]$$

which prove the case where $(f_p \, \llcorner \, b)_1$ is defined.

Now, assume that $f_p \, \llcorner \, b$ is not general of order 1. Then

$$(\mathbf{\omega}_p \, \llcorner \, \beta) \wedge (\mathbf{\omega}_p \, \llcorner \, \beta)' \equiv 0 \quad \text{on } U_\beta$$

Hence

$$0 \equiv [(\mathbf{\omega}_p \, \llcorner \, \beta) \wedge (\mathbf{\omega}_{p-1} \wedge \mathbf{\omega}^{(p+1)} \, \llcorner \, \beta)] \, \llcorner \, \alpha = (\mathbf{\omega}_{p-1} ; \beta)(\mathbf{\omega}_{p+1} \llcorner \beta \wedge \alpha)$$

Therefore $||f_{p-1};b||^2\ ||f_{p+1};b||^2 \equiv 0$ and the statement is trivially true; q.e.d.

Lemma 9.4. Let p be an integer with $0 \leqq p \leqq n-1$. Take $a \in G_p(V)$. Then

$$I(\log||x:a||^{-2}) = \int_{\mathbb{P}(V^*)} \log ||x:a||^{-2}\, \ddot{\omega}_0^n = \sum_{\nu=n-p}^{n} \frac{1}{\nu}.$$

Proof. (Compare Weyl [48] §5, p. 147 or [36] Hilfsatz 1 p. 107). An orthonormal base $\alpha_0,\ldots,\alpha_n$ of V^* exists such that $\mathbb{P}(\alpha_0 \wedge \ldots \wedge \alpha_p) = a$. Take $\xi \in \mathbb{P}^{-1}(x)$. Then $\xi = x_0\alpha_0 + \ldots + x_n\alpha_n$. Hence

$$||x:a||^2 = \frac{|\alpha_0 \wedge \ldots \wedge \alpha_p \wedge \xi|^2}{|\alpha_0 \wedge \ldots \wedge \alpha_p|^2 |\xi|^2} = \frac{|x_{p+1}|^2 + \ldots + |x_n|^2}{|x_0|^2 + \ldots + |x_n|^2}$$

Define $I = I(\log ||x:a||^{-2})$. Then

$$I = \frac{1}{(n+1)!} \int_{V^*} e^{-|x_0|^2 - \ldots - |x_n|^2} \log\frac{|x_0|^2 + \ldots + |x_n|^2}{|x_{p+1}|^2 + \ldots + |x_n|^2}\, \upsilon^{n+1}$$

$$= \int_0^\infty \ldots \int_0^\infty e^{-t_0 - \ldots - t_n} \log\frac{t_0 + \ldots + t_n}{t_{p+1} + \ldots + t_n}\, dt_0 \ldots dt_n$$

$$= \sum_{q=0}^{p} \int_0^\infty \ldots \int_0^\infty e^{-t_q - \ldots - t_n} \log\frac{t_q + \ldots + t_n}{t_{q+1} + \ldots + t_n}\, dt_q \ldots dt_n$$

Define

$$t_q = \tau(s_{q+1} + \ldots + s_n) \qquad 0 < \tau < 1 \qquad 0 < s_\nu$$

$$t_\nu = (1-\tau)\ s_\nu \qquad \nu = q+1,\ldots,n$$

$$t_q +\ldots+ t_n = s_{q+1} +\ldots+ s_n$$

$$\tau = \frac{t_q}{t_q +\ldots+ t_n} \qquad 1 - \tau = \frac{t_{q+1} +\ldots+ t_n}{t_q +\ldots+ t_n}$$

$$s_\nu = t_\nu \frac{t_{q-1} +\ldots+ t_n}{t_{q+1} +\ldots+ t_n}$$

$$\frac{\partial(t_q,\ldots, t_n)}{\partial(\tau, s_{q+1},\ldots,s_n)} = (1-\tau)^{n-1}\ (s_{q+1} +\ldots+ s_n)$$

Abbreviate $S(q) = s_{q+1} +\ldots+ s_n$

$$I = \sum_{q=0}^{p} \int_0^1 \int_0^\infty \ldots \int_0^\infty e^{-S(q)} \log \frac{1}{1-\tau}(1-\tau)^{n-q-1} S(q)\, d\tau\, ds_{q+1}\ldots ds_n$$

$$= \sum_{q=0}^{p} (n-q) \int_0^1 \tau^{n-q-1} \log \frac{1}{\tau}\ d\tau$$

$$= \sum_{q=0}^{p} \int_0^1 \tau^{n-q} \frac{1}{\tau}\, d\tau = \sum_{q=0}^{p} \frac{1}{n-q} = \sum_{\nu=n-p}^{n} \frac{1}{\nu}\ ,$$

q.e.d.

<u>Lemma 9.5</u>. Take $y \in G_p(V)$ and $b \in G_q(V^*)$ with $0 \leqq q < p < n$. Assume that $y \llcorner b$ exists. Then

$$I(\log \|y \llcorner b; x\|^{-2}) = \sum_{\nu=p-q}^{n} \frac{1}{\nu}\ .$$

Proof. Take $\mathfrak{y} \in \mathbb{P}^{-1}(y)$ and $\beta \in \mathbb{P}^{-1}(b)$ and $\xi \in \mathbb{P}^{-1}(x)$. Let $\mathfrak{v}$ be an orthonormal base of V. Let ε be the dual base. Consider the dual maps $D_{\mathfrak{v}}$ and D_ε. Then

$$D_{\mathfrak{v}} \mathfrak{y} = \eta \in \tilde{G}_{n-p-1}(V^*)$$

$$\gamma = \eta \wedge \beta \in \tilde{G}_{n-p+q}(V^*)$$

$$D_\varepsilon \gamma = (D_\varepsilon \eta) \llcorner \beta = (-1)^{(p+1)(n-p)} \quad \mathfrak{y} \llcorner \beta \neq 0$$

$$D_\varepsilon(\gamma \wedge \xi) = D_\varepsilon \gamma \llcorner \xi = (-1)^{(p+1)(n-p)} \; (\mathfrak{y} \llcorner \beta) \llcorner \xi$$

$$= (-1)^{(p+1)(n-p)} \quad \mathfrak{y} \llcorner \beta \wedge \xi$$

Define $c = \mathbb{P}(\gamma) \in G_{n-p+q}(V^*)$. Then

$$||y \llcorner b;x|| = \frac{|\mathfrak{y} \llcorner \beta \wedge \xi|}{|\mathfrak{y} \llcorner \beta| \, |\xi|} = \frac{|D_\varepsilon(\gamma \wedge \xi)|}{|D_\varepsilon \gamma| \; |\xi|}$$

$$= \frac{|\gamma \wedge \xi|}{|\gamma| \; |\xi|} = ||c:x||$$

Hence

$$I\ (\log ||y \llcorner b;x||^{-2}) = I\ (\log ||c:x||^{-2}) = \sum_{\nu=p-q}^{n} \frac{1}{\nu}$$

q.e.d.

<u>Lemma 9.6</u>. Let p be an integer with $0 \leqq p < n$. Take $y \in G_p(V)$. Then

$$I\ (\log ||y;x||^{-2}) = \sum_{\nu=p+1}^{n} \frac{1}{\nu}$$

Proof. Define $\mathfrak{y}$, $\mathfrak{v}$, η, ε as in the proof of Lemma 9.5.

Define $c = \mathbb{P}(\eta) \in G_{n-p-1}(V^*)$. Then

$$D_\varepsilon(\eta \wedge \xi) = (D_\varepsilon D_{\mathfrak{n}} \mathfrak{y}) \llcorner \xi = (-1)^{(p+1)(n-p)} \mathfrak{y} \llcorner \xi$$

$$||y;x|| = \frac{|\mathfrak{y} \llcorner \xi|}{|\mathfrak{y}| \; |\xi|} = \frac{|D_\varepsilon(\eta \wedge \xi)|}{|D_\varepsilon \eta| \; |\xi|} = \frac{|\eta \wedge \xi|}{|\eta| \; |\xi|} = ||y:x||$$

Hence

$$I\ (\log ||y;x||^{-2}) = I\ (\log ||c:x||^{-2}) = \sum_{\nu=p+1}^{n} \frac{1}{\nu}$$

q.e.d.

Lemma 9.7. Assume (A1) and (A3). Let p and q be integers with $0 \leqq q + 1 < p \leqq n$. Let $\mathfrak{w} : M \to \tilde{G}_p(V)$ be a holomorphic map. Take $\beta \in \tilde{G}_q(V^*)$ if $q \geqq 0$. Define $\mathfrak{y} = \mathfrak{w} \llcorner \beta$ if $q \geqq 0$ and $\mathfrak{y} = \mathfrak{w}$ if $q = -1$. Assume $\mathfrak{y}(z) \neq 0$ for all $z \in M$. For $\xi \in V^*$, define

$$L_\xi = \{z \in M \mid \mathfrak{y}(z) \llcorner \xi = 0\}$$

Then the set

$$E = \{\xi \in V^* \mid \dim L_\xi \geqq m-1\}$$

is almost thin in V^*. Also $\mathbb{P}(E)$ is almost thin in $\mathbb{P}(V^*)$.

Proof. The set

$$N = \{(z,\xi) \in M \times V^* \mid \mathfrak{y}(z) \llcorner \xi = 0\}$$

is analytic in $M \times V^*$. Let $\pi: N \to M$ and $\rho: N \to V^*$ be the projection. Define the rank and pseudo-rank as in [2]. Take any $z \in M$. Write $\beta = \beta_0 \wedge \cdots \wedge \beta_q$. Let $\mathfrak{v} = (\mathfrak{v}_0, \ldots, \mathfrak{v}_n)$ be a base of V and let ε be the dual base. Define $D_{\mathfrak{v}}(\mathfrak{w}(z)) = \alpha$. Write $\alpha = \beta_{q+1} \wedge \cdots \wedge \beta_{q+n-p}$. If $s_0, s_1, \ldots,$ denote +1 or -1, then

$$D_\varepsilon(\beta_0 \wedge \cdots \wedge \beta_{q+n-p}) = s_0\, D_\varepsilon(\alpha \wedge \beta) = s_0\, (D_\varepsilon \alpha) \llcorner \beta$$

$$= s_1\, \mathfrak{w}(z) \llcorner \beta = s_1\, \mathfrak{y}(z) \neq 0.$$

Hence, a base $\beta_0, \ldots, \beta_n$ exists. Then $\xi = \xi_0 \beta_0 + \cdots + \xi_n \beta_n$ and

$$\mathfrak{y}(z) \llcorner \xi = (\mathfrak{w}(z) \llcorner \beta) \llcorner \xi = \mathfrak{w}(z) \llcorner \beta \wedge \xi = s_2\, D_\varepsilon(\alpha) \llcorner \beta \wedge \xi$$

$$= s_2\, D_\varepsilon(\alpha \wedge \beta \wedge \xi)$$

$$= s_2\, D_\varepsilon\Big(\sum_{\nu=n-p+q+1}^{n} \beta_0 \wedge \cdots \wedge \beta_{n-p+q} \wedge \beta_\nu\, \xi_\nu\Big)$$

Hence $\pi^{-1}(z) = \{z\} \times \{\xi \mid \mathfrak{y}(z) \llcorner \xi\}$ has dimension n-p+q+1. If $(z, \xi) \in N$, then

$$\dim_{(z,\xi)} N = \widetilde{\operatorname{rank}}_{(z,\xi)} \pi^{-1}(z) + \dim \pi^{-1}(z) \leqq m+n-p+q+1$$

$$\leqq m+n-1-(p-q-2) \leqq m+n-1.$$

Let $\{N_\lambda\}_{\lambda \in \Lambda}$ be the family of branches of N. Define $\rho_\lambda = \rho | N_\lambda$. Then

$$D_\lambda = \{(z,\xi) \in N_\lambda \mid \text{rank}_{(z,\xi)} \, \rho_\lambda \leqq n\}$$

is an analytic subset of N_λ and $D'_\lambda = \rho_\lambda(D_\lambda)$ is almost thin in V^*. ([2] Lemma 1.30.). Hence $D' = \bigcup_{\lambda\in\Lambda} D'_\lambda$ is almost thin in V^*. Take $\xi \in V^* - D'$. Then $\rho^{-1}(\xi) = L_\xi \times \{\xi\}$. If $\rho^{-1}(\xi) = \emptyset$, then $\dim L_\xi = -1 < m-1$. Hence $\xi \in V^* - E$. If $\rho^{-1}(\xi) \neq \emptyset$, take any $z \in L_\xi$. Then $(z,\xi) \in N_\lambda - D_\lambda$ for some $\lambda \in \Lambda$. Hence $\text{rank}_{(z,\xi)}\rho_\lambda = n+1$ and

$$\dim_z L_\xi = \dim N_\lambda - \text{rank}_{(z,\xi)}\rho_\lambda \leqq m-2.$$

Therefore $\xi \in V^* - E$. Hence $E \subseteq D'$ is almost thin. If $\xi \in E$ and $0 \neq z \in \mathbb{C}$, then $z\xi \in E$. Hence $\mathbb{P}(E)$ almost thin in $\mathbb{P}(V^*)$;

q.e.d.

Theorem 9.8. Assume (A1) - (A6). Let p and q be integers with $0 \leqq q+1 < p \leqq n$. If $q \geqq 0$, take $b \in G_q(V^*)$ and assume that f_p is adjusted to b. Let (G,g,ψ) be a bump on M. Then

$$I(T_p(G,b \wedge x)) = T_p(G;b)$$

$$I(T_p(G;x)) = T_{f_p}(G)$$

Proof. Take a positive form η of bidegree (m,m) and class C^∞ on M. Then a non-negative function F is almost everywhere defined on M by

$$(f_b \llcorner b)^* (\ddot{\omega}_{p-q-1}) \wedge \chi = F\,\eta$$

If $x \in \mathbb{P}(V^*)$ such that $b \wedge x$ exists and if f_p is adjusted to $b \wedge x$, define a non-negative function $F[x]$ almost everywhere on M by

$$(f_p \llcorner b \wedge x)^* (\ddot{\omega}_{p-q-2}) \wedge \chi = F[x]\ \eta$$

Now, $I(F[x]) = F$ almost everywhere on M is claimed.

Take $z_0 \in M$. Take $\alpha \in \mathfrak{T}_M$ with $z \in U_\alpha$ such that an irreducible representation $\mathfrak{v} : U_\alpha \to V$ of f on U_α exists. Write $\mathfrak{w} = \mathfrak{v}_{p\alpha}: U_\alpha \to \underset{p+1}{\wedge} V$. If $q = -1$, define $\mathfrak{y} = \mathfrak{w}$. If $q \geqq 0$, take $\beta \in P^{-1}(b)$ with $|\beta| = 1$ and define $\quad = \quad {}_p \quad \beta$. Then is a representation of f_p b (resp. f_p if $q = -1$) on U_α. The open set $U = U_\alpha - \mathfrak{y}^{-1}(0)$ is dense in U_α. Let $H \neq \emptyset$ be any open set such that $\overline{H}$ is compact and contained in U and such H has a boundary manifold $dH = \partial H$. Let $\rho \geqq 0$ be any function of class C^∞ on M with compact support in H. Then $(H, \emptyset, \rho)$ is a bump on M, and $\mathfrak{y}$ is a simple representation of $f_p \llcorner b$ on U. (If $q = -1$, read $f_p \llcorner b = f_p$)

$$\int_H \rho\ I(F[x])\eta = \int_H \rho\ F\eta + \sum_{\nu=p-q}^{n} \frac{1}{\nu} \frac{1}{4\pi} \int_H dd^\perp \rho \wedge \chi$$

Hence $\int_H dd^\perp \rho \wedge \chi = \int_{dH} d^\perp \rho \wedge \chi = 0$. Hence

$$\int_H \rho\ I(F[x])\ \eta = \int_H \rho\ F\ \eta < +\infty$$

for all $\rho \geqq 0$ of class C^∞ with compact support in H. Therefore $I(F[x])$ exists almost everywhere on H, hence on U and therefore, almost everywhere on M. Also $I(F[x]) = F$ almost everywhere on M. Consequently

$$I(T_p(G; b \wedge x)) = I(\int_G \psi\, F[x]\, \chi) =$$

$$= \int_G \psi\, I(F[x])\, \chi = \int_G \psi\, F\, \chi = T_p(G;x)$$

q.e.d.

Lemma 9.9. Assume (A3). Let p and q be integers with $0 \leqq q < p-1 < n-1$. Take $y \in G_p(V)$ and $a \in \mathbb{P}(V^*)$ and $b \in G_q(V^*)$. Assume that $b \wedge a$ and $y \llcorner b \wedge a$ exist. Then

$$I(\log \|(y \llcorner b \wedge x);a\|^2) = \log \|y \llcorner b;a\|^2 - \frac{1}{p-q-1}.$$

Proof. Take $\mathfrak{y} \in \mathbb{P}^{-1}(y)$ and $\beta \in \mathbb{P}^{-1}(b)$ and $\xi \in \mathbb{P}^{-1}(x)$ and $\alpha \in \mathbb{P}^{-1}(a)$. Then

$$\|y \llcorner b \wedge x;a\| = \frac{|(\mathfrak{y} \llcorner \beta \wedge \xi) \llcorner \alpha|}{|\mathfrak{y} \llcorner \beta \wedge \xi|\ |\alpha|}$$

$$= \frac{|\mathfrak{y} \llcorner \beta \wedge \alpha \wedge \xi|}{|\mathfrak{y} \llcorner \beta \wedge \alpha|\,|\xi|} \ \frac{|\mathfrak{y} \llcorner \beta \wedge \alpha|}{|\mathfrak{y} \llcorner \beta|\,|\alpha|} \ \frac{|\mathfrak{y} \llcorner \beta|\ |\xi|}{|\mathfrak{y} \llcorner \beta \wedge \xi|}$$

$$= \frac{\|y \llcorner b \wedge a;x\|\ \ \|y \llcorner b;a\|}{\|y \llcorner b;x\|}$$

Hence

$$I(\log \|y \llcorner b \wedge x;a\|^2) = \log \|y \llcorner b;a\|^2 - \sum_{\nu=p-q-1}^{n} \frac{1}{\nu} + \sum_{\nu=p-q}^{n} \frac{1}{\nu}$$

$$= \log \|y \llcorner b;a\|^2 - \frac{1}{p-q-1}$$

q.e.d.

Lemma 9.10. Assume (A3). Let p be an integer with $0 < p < n$. Take $y \in G_p(V)$ and $a \in \mathbb{P}(V^*)$ such that $y \llcorner a$ exists. Then

$$I\,(\log \|y \llcorner x;a\|^2) = \log \|y;a\|^2 - \frac{1}{p}$$

The proof is the same as for Lemma 9.9 if $b \wedge x$, $\beta \wedge \xi$, $\beta \wedge \alpha$, $\mathfrak{y} \llcorner \beta$, $b \wedge a$, $y \llcorner b$ read $x, \xi, \alpha, \mathfrak{y}, a, y$ respectively.

Lemma 9.11. Assume (A3). Let p and q be integers with $0 \leqq q < p < n$. Take $y \in G_p(V)$ and $b \in G_q(V^*)$ such that $y \quad b$ exists. Then

$$I(\log \|y;b \wedge x\|^2) = \log \|y;b\|^2 - \sum_{\nu=p}^{n-1} \frac{1}{\nu - q}$$

Proof. Take $\mathfrak{y} \in \mathbb{P}^{-1}(y)$ and $\beta \in \mathbb{P}^{-1}(b)$ and $\xi \in \mathbb{P}^{-1}(x)$. Then

$$\|y;b \wedge x\| = \frac{|\mathfrak{y} \llcorner \beta \wedge \xi|}{|\mathfrak{y}|\;|\beta \wedge \xi|}$$

$$= \frac{|(\mathfrak{y} \llcorner \beta) \llcorner \xi|}{|\mathfrak{y} \llcorner \beta|\;|\xi|} \quad \frac{|\mathfrak{y} \llcorner \beta|}{|\mathfrak{y}|\;|\beta|} \quad \frac{|\beta|\,|\xi|}{|\beta \wedge \xi|}$$

$$= \frac{\|y \llcorner b;x\|\;\|y;b\|}{\|b:x\|}$$

Hence

$$I\ (\log\ ||y;b \wedge x||^2) = \log\ ||y;b||^2 - \sum_{\nu=p-q}^{n} \frac{1}{\nu} - \sum_{\nu=n-q}^{n} \frac{1}{\nu}$$

$$= \log\ ||y;b||^2 - \sum_{\nu=p}^{n-1} \frac{1}{\nu-q}$$

q.e.d.

Lemma 9.12. Assume (A3). Let q be an integer with $0 \leqq q < n-1$. Take $b \in G_q(V^*)$ and $a \in \mathbb{P}(V^*)$ such that $b \wedge a$ exists. Then

$$I\ (\log\ ||b \wedge x:a||^2) = \log\ ||b:a||^2 - \frac{1}{n-1-q}$$

Proof. Take $\beta \in \mathbb{P}^{-1}(b)$ and $\xi \in \mathbb{P}^{-1}(x)$ and $\alpha \in \mathbb{P}^{-1}(a)$. Then

$$||b \wedge x:a|| = \frac{|\beta \wedge \xi \wedge \alpha|}{|\beta \wedge \xi||\alpha|}$$

$$= \frac{|\beta \wedge \alpha \wedge \xi|}{|\beta \wedge \alpha||\xi|} \quad \frac{|\beta||\xi|}{|\beta \wedge \xi|} \quad \frac{|\beta \wedge \alpha|}{|\beta||\alpha|}$$

$$= \frac{||b \wedge a:x|| \quad ||b:a||}{||b:x||}$$

Hence

$$I(\log\ ||b \wedge x:a||^2) = \log\ ||a:b||^2 - \sum_{\nu=n-q-1}^{n} \frac{1}{\nu} + \sum_{\nu=n-q}^{n} \frac{1}{\nu}$$

$$= \log\ ||a:b||^2 - \frac{1}{n-q-1}$$

q.e.d.

For $0 \leqq q < p < n$ if $0 < \lambda < 1$ and if

$w \in G_{p-1}(V)$ $y \in G_p(V)$ $z \in G_{p+1}(V)$

$a \in \mathbb{P}(V^*)$ $b \in G_q(V^*)$

such that $y \llcorner b$ and $y \llcorner b \wedge a$ exist define

$$F_q(y,w,z,a,b) = \frac{||b:a||^2\ ||w;b||^2\ ||z;b \wedge a||^2}{||y \llcorner b;a||^{2\lambda}\ ||y;b||^4}$$

If $0 \leqq p < n$ and $(y,z,a) \in G_p(V) \times G_{p+1}(V) \times \mathbb{P}(V^*)$ such that y a exists, define

$$F(y,z,a) = \frac{||z;a||^2}{||y;a||^2}$$

<u>Lemma 9.13</u>. Assume (A3). Then

$$e\ I(F_{q+1}(y,w,z,a,b \wedge x)) \geqq F_q(y,z,w,a,b)$$

$$e\ I(F_0(y,w,z,a,x)) \geqq F(y,z,a).$$

Proof. At first consider the case $q \geqq 0$. Then

$$\log I(F_{q+1}(y,w,z,a,b \wedge x))$$

$$\geqq I\ (\log F_{q+1}(y,w,z,a,b \wedge x))$$

$$\geqq I\ (\log ||b \wedge x:a||^2) + I\ (\log||w;b \wedge x||^2)$$

$$+ I(\log||z;b \wedge a \wedge x||^2) - 2I(\log||y;b \wedge x||^2) - \lambda I(\log||y \llcorner b \wedge x;a||^2)$$

$$= \log ||b:a||^2 - \frac{1}{n-q-1} + \log ||w;b||^2 - \sum_{\nu=p-1}^{n-1} \frac{1}{\nu-q}$$

$$+ \log ||z;b \wedge a||^2 - \sum_{\nu=p+1}^{n-1} \frac{1}{\nu-q-1} - 2 \log ||y;b||^2$$

$$+ 2\sum_{\nu=p}^{n-1} \frac{1}{\nu-q} - \lambda\, I\, (\log ||y \quad b;a||^2) + \frac{\lambda}{p-q-1}$$

$$= \log F_q(y,w,z,a,b) - \frac{1-\lambda}{p-q-1}$$

$$\geqq \log F_q(y,w,z,a,b) - 1.$$

Hence

$$I\, (F_{q+1}(y,w,z,a,b)) \geqq e\, F_q(y,w,z,a,b).$$

If $q = -1$, then

$$\log I\, (F_0(y,w,z,a,x)) \geqq I\, (\log F_0(y,w,z,a,x))$$

$$= I\, (\log ||a:x||^2) + I(\log ||w;x||^2) + I(\log ||z;a \wedge x||^2)$$

$$- 2I\, (\log ||y;x||^2) - \lambda I\, (\log ||y \llcorner x;a||^2)$$

$$= - \frac{1}{n} - \sum_{\nu=p}^{n} \frac{1}{\nu} + \log ||z;a||^2 - \sum_{\nu=p+1}^{n-1} \frac{1}{\nu} + 2 \sum_{\nu=p+1}^{n} \frac{1}{\nu}$$

$$- \lambda \log ||y;a||^2 + \frac{\lambda}{p}$$

$$= \log F(y,z,a) - \frac{1-\lambda}{p}$$

$$\geqq \log F(y,z,a) - 1.$$

q.e.d.

Theorem 9.14. Assume (A1) - (A7). Let p and q be integers with $-1 \leqq q < p < n$. Take $a \in \mathbb{P}(V^*)$. If $q = -1$, assume that f_p is adjusted to a. If $q \geqq 0$, take $b \in G_q(V^*)$ such that $b \wedge a$ exists and such that f_p is adjusted to $b \wedge a$. Take $0 < \lambda < 1$. Let (G,g,ψ) be a bump on M. If $q \geqq 0$, then

$$\int_G \psi\, F_q(f_p, f_{p-1}, f_{p+1}, a, b)\, H_p \leqq \leqq \frac{e^{p-q-1}}{(1-\lambda)^2}\, c_n\, [T_p(G;b) + \Phi^+(G)] \tag{9.1}$$

If $q = -1$, then

$$\int_G \psi\, F(f_p, f_{p+1}; a) H_p \leqq \frac{e^p}{(1-\lambda)^2}\, c_n\, [T_{f_p}(G) + \Phi^+(G)] \tag{9.2}$$

Here $c_n > 2$ is a constant which depends on n only.

Proof. If $p = 0$, then $q = -1$ and $F(f_0, f_1, a) = \dfrac{||f_1;a||^2}{||f,a||^{2\lambda}}$ and Proposition 9.1 implies Theorem 9.14. Assume $p > 0$. If $q = p-1$, this is Theorem 9.3. Assume that the statement is correct for q+1 with $0 \leqq q+1 < p$. Then prove it for q. By assumption

$$\int_G \psi\, F_{q+1}(f_p, f_{p-1}, f_{p+1}, a, b \wedge x)$$

$$\leqq \frac{e^{p-q-2}}{(1-\lambda)^2}\, c_n\, [T_p(G, b \wedge x) + \Phi^+(G)]$$

if $q \geqq 0$. Apply the operator I. Lemma 9.13 and 9.8 imply the assertion (9.1) for $q \geqq 0$. If $q = -1$, replace $F_{q+1}(f_p, f_{p-1}, f_{p+1}, a, b \wedge x)$ by $F_0(F_p, f_{p-1}, f_{p+1}, a, x)$ and $f_p \llcorner b \wedge x$ by $f_p \llcorner x$. Then (9.1) is obtained, q.e.d.

Lemma 9.15. Assume (A1) - (A6). Take $a \in \mathbb{P}(V^*)$. Then f_p is adjusted to a for all $p = 0,1,\ldots,n$.

Proof. Take $\alpha \in \mathbb{P}^{-1}(a)$. Take $\beta \in \mathfrak{S}_M$ such that an irreducible representation $\mathfrak{w} : U_\beta \to V$ of f on U_β exists. By (A6), the map f is general for B. Hence f is generic. Therefore f is adapted to a. Hence $(\mathfrak{w};\alpha) \not\equiv 0$ on U_β. Take $x \in U_\beta$ with $\mathfrak{w}_{n\beta}(x) \neq 0$ and $(\mathfrak{w}(x);\alpha) \neq 0$. Define $\mathfrak{v}_\nu = \mathfrak{w}_\beta^{(\nu)}(x)$. Then $\mathfrak{v}_0, \ldots, \mathfrak{v}_n$ is a base of V. Let $\varepsilon_0, \varepsilon_1, \ldots, \varepsilon_n$ be the dual base. Then

$$\alpha = \alpha_0\varepsilon_0 + \ldots + \alpha_n\varepsilon_n$$

$$\alpha_0 = (\mathfrak{v}_0;\alpha) = (\mathfrak{w}(x);\alpha) \neq 0$$

If $0 \leqq p \leqq n$, then

$$\mathfrak{w}_{p\beta}(x) \llcorner \alpha = (\mathfrak{v}_0 \wedge \ldots \wedge \mathfrak{v}_p) \llcorner (\alpha_0\varepsilon_0 + \ldots + \alpha_n\varepsilon_n)$$

$$= \sum_{\nu=0}^{p} (-1)^{\nu-1} \alpha_\nu \; \mathfrak{v}_0 \wedge \ldots \wedge \mathfrak{v}_{\nu-1} \wedge \mathfrak{v}_{\nu+1} \wedge \ldots \wedge \mathfrak{v}_n \neq 0$$

because $\alpha_0 \neq 0$. Hence $\mathfrak{w}_{p\beta} \llcorner \alpha \not\equiv 0$; q.e.d.

Theorem 9.16. Ahlfors estimate.

Assume (A1) - (A7). Let p be an integer with $0 \leqq p < n$. Let

$0 < \lambda < 1$. Take $a \in \mathbb{P}(V^*)$. Then

$$\int_G \psi \frac{||f_{p+1};a||^2}{||f_p;a||^{2\lambda}} H_p \leqq \frac{c_n e^p}{(1-\lambda)^2} [T_{f_p}(G) + \Phi^+(G)]$$

where the constant $c_n > 2$ depends on n only and increases with n. Here $||f_n;a|| \equiv 1$.

Proof. By Lemma 9.15, the meromorphic map f_p is adjusted to a. Hence (9.2) holds. q.e.d.

For general reference to this section see also Ahlfors [1] and Wu [51].

X. PSEUDO-CONVEX EXHAUSTION

The main purpose of this investigation is to find estimates of the deficit function and to apply these estimates to the Bezout problem. Therefore, it become imperative to select more concrete objects for $(G,g,\psi),\chi$ and B. There are two basically different methods. The first is to solve the Dirichlet problem $dd^{\perp}\psi \wedge \chi \equiv 0$ on $G-\overline{g}$ such that $\psi|\partial G = 0$ and $\psi|\partial g = R > 0$ are constant with $\frac{1}{2\pi}\int_{\partial g} d^{\perp}\psi \wedge \chi = 1$. Then $D_f(G,a) \equiv 0$ and the deficit does not have to be estimated. This method was used in Weyl [48] and [36]. The first and second main theorems were derived. Unfortunately all the information on the complex analytic properties of M and are packed into ψ. If $m > 1$, practically nothing is known about the behavior of ψ. Except the situation of [35], the application to the Bezout problem seems to be hopeless at the present time.

The second method uses an exhaustion function. Since the Bezout problem gives a pseudo-convex exhaustion function the second method shall be investigated here.

Let M be a connected, non-compact, complex manifold of dimension $m > 0$. A non-negative function $\tau: M \to \mathbb{R}[0,+\infty)$ of class C^∞ is said to be an exhaustion function if τ is proper and if each $\tau^{-1}(r)$ has measure zero on M. Define

$$G_r = G_r(\tau) = \{z \in M \mid \tau(z) < r\}$$

$$\Gamma_r = \tau^{-1}(r) = \{z \in M \mid \tau(z) = r\}$$

Then $\overline{G}_r$ and $G_r \cup \Gamma_r$ are compact with

$$dG_r \subseteq \partial G_r \subseteq \Gamma_r$$

By Sards theorem $\Gamma_r = dG_r$ for almost all $r \geqq 0$. Let E_τ be the set of all $r \geqq 0$ such that $\Gamma_r - dG_r$ has $\mathfrak{H}^{2m-1}$-measure. Obviously, $\mathbb{R}[0,+\infty)-E_\tau$ has measure zero. Observe that $G_0 = \emptyset$. Assume $0 \in E_\tau$. If $r \geqq 0$, write $\int_{dG_r} = \int_{\Gamma_r}$. If $r \in E_\tau$, then G_r is a Stokes domain. If τ is real analytic, then $E_\tau = \mathbb{R}[0,+\infty)$.

The exhaustion τ is said to be pseudo-convex if $dd^c\tau \geqq 0$ on M and if $dd^c\tau > 0$ almost everywhere on M. The exhaustion τ is said to be strictly pseudo-convex if $dd^c\tau > 0$

everywhere on M. Observe that M is a Stein manifold if and only if a strictly pseudo-convex exhaustion exists. A complex manifold M is said to be pseudo-convex if there exists a pseudo-convex exhaustion of M. Every modification of a Stein manifold is pseudo-convex.

Now, the old general assumptions (A_x) will be replaced by new general assumptions (B_x).

(B1) Let M be a non-compact, connected complex manifold of dimension $m > 0$.

(B2) Let τ be a pseudo-convex exhaustion of M. Define $r_0 = \inf \{r \in \mathbb{R}[0,+\infty) \mid G_r \neq \emptyset\}$.

(B3) Let V be a hermitian vector space of dimension $n+1$ with $n > 0$.

(B4) Let $f: M \longrightarrow \mathbb{P}(V)$ be a meromorphic map.

Observe that $G_r = \bigcup_{t<r} G_t$ and $G_r \cup \Gamma_r = \bigcap_{t>r} G_t$. Because Γ_r has measure zero, the following Lemma holds.

Lemma 10.1. Assume (B1) and (B2). Let ξ be a locally integrable differentiable form of degree 2m on M. Then the integral

$$\int_{G_r} \xi = \int_{G_r \cup \Gamma_r} \xi$$

exists for each $r \geqq 0$ and is a continuous function of r.

Define

$$\varphi = \frac{1}{2\pi} dd^c \tau \geqq 0$$

on M. The <u>volume of G_r</u> in respect to φ is given by

$$\Phi(r) = \int_{G_r} \varphi^m$$

Then Φ is a continuous increasing function with $\Phi(r) > 0$ if $r > r_0$.

A function $u: \mathbb{R}[0,+\infty) \to \mathbb{R}[0,+\infty)$ is said to be a <u>change of scale for τ</u> if and only if the following conditions are satisfied

1) The function u is of class C^∞.
2) If $x \geqq 0$, then $u(x) \geqq 0$ and $u'(x) > 0$.
3) Also $u(x) \to +\infty$ for $x \to +\infty$.
4) $d^\perp d(u \circ \tau) \geqq 0$ on M and $d^\perp d(u \circ \tau) > 0$ almost everywhere on M.

Obviously, $u \circ \tau$ is again a pseudo-convex exhaustion function.

A function $u: \mathbb{R}[0,\infty) \to \mathbb{R}[0,\infty)$ of class C^∞ is said to be a <u>convex change of scale</u> if and only if $u(x) \to \infty$ for $x \to +\infty$ and if $u(x) \geqq 0$ and $u'(x) > 0$ and $u''(x) \geqq 0$ for all $x \geqq 0$.

<u>Lemma 10.2</u>. Assume (B1) and (B2). A convex change of scale is a change of scale for τ.

Proof. If u is a convex change of scale, then

$$dd^c(u \circ \tau) = u' \circ \tau \, dd^c\tau + (u'' \circ \tau)\, d\tau \wedge d^c\tau$$

Here

$$d\tau \wedge d^c\tau = 2i\, \partial\tau \wedge \partial\bar{\tau} \geqq 0$$

Hence $dd^c(u \circ \tau) \geqq u' \circ \tau\, dd^c\tau$. Because $u' \circ \tau > 0$, the form $dd^c\, u \circ \tau$ is non-negative and almost everywhere positive; q.e.d.

The set of all convex change of scales is denoted by $\mathfrak{U}$. The set of all changes of scales for τ is denoted by $\mathfrak{U}_\tau$. If $u \in \mathfrak{U}_\tau$ define

$$\varphi_u = \frac{1}{2\pi}\, dd^c\, (u \circ \tau)$$

Then

$$\varphi_u^q = (u' \circ \tau)\varphi^q + q(u'' \circ \tau)(u' \circ \tau)^{q-1} d\tau \wedge d^c\tau \wedge \varphi^{q-1} \quad (10.1)$$

for $q \geqq 1$. Also

$$d^c(u \circ \tau) = u' \circ \tau\ \ d^c\tau \quad (10.2)$$

$$d^c(u \circ \tau) \wedge \varphi_u^q = (u' \circ \tau)^{q+1}\, d^c\tau \wedge \varphi^q \quad (10.3)$$

for $q \geqq 0$. Define

$$\Phi_u(r) = \int_{G_r} \varphi_u^m$$

Then Φ_u is an increasing, continuous function with $\Phi_u(r) > 0$ if $r > r_0$.

Lemma 10.3. Assume (B1) and (B2). Let u be a change of scale for τ. Then

$$\Phi_u(r) = (u'(r))^m \, \Phi(r)$$

Proof. By continuity, it suffices to prove the statement for $r \in E_\tau$. Then

$$\Phi_u(r) = \frac{1}{2\pi}\int_{G_r} d(d^c u \circ \tau \wedge \varphi_u^{m-1}) = -\frac{1}{2\pi}\int_{G_r} d(u' \circ \tau)^m d^c\tau \wedge \varphi^{m-1})$$

$$= (u'(r))^m \frac{1}{2\pi}\int_{\Gamma_r} d^c\tau \wedge \varphi^{m-1} = (u'(r))^m \int_{G_r} \varphi^m = (u'(r))^m \Phi(r)$$

q.e.d.

If $u \in \mathfrak{U}_\tau$, define

$$\eta_u(r) = \int_0^r \frac{dt}{(u'(t))^{m-1}} \geqq 0$$

Then η_u is an increasing function of class C^∞ with $\eta_u' = (u')^{1-m}$. Also $\eta_{Id} = Id$. If $0 \leqq s < r < +\infty$, define

$$\psi_{u,r,s}(z) = \begin{cases} 0 & \text{if } \tau(z) \geqq r \\ \eta_u(r) - \eta_u(\tau(z)) & \text{if } s \leqq \tau(z) < r \\ \eta_u(r) - \eta_u(s) & \text{if } \tau(z) < s \end{cases}$$

$$\psi_{r,s}(z) = \psi_{Id,r,s}(z) = \begin{cases} 0 & \text{if } z \in M - G_r \\ r - \tau(z) & \text{if } z \in G_r - G_s \\ r - s & \text{if } z \in G_s \end{cases}$$

The function $\psi_{u,r,s}$ is non-negative and continuous. If $r_0 < r \in E_\tau$ and $s \in E_\tau$ with $0 \leqq s < r < +\infty$, then $(G_r, G_s, \psi_{u,r,s})$ is a bump on M. If $r > r_0$, then $(G_r, \emptyset, \psi_{r0})$ is a bump on M where s = 0. Define

$$A_f(r) = \int_{G_r} f^*(\ddot{\omega}_0) \wedge \varphi^{m-1} \geqq 0 \qquad \text{for } r \geqq 0$$

$$T_f(r,s) = \int_s^r A_f(t)\, dt \geqq 0 \qquad \text{for } 0 \leqq s < r$$

$$T_f(r) = T_f(r,0) = \int_0^r A_f(t)\, dt \geqq 0 \qquad \text{for } r \geqq 0$$

Then A_f is a continuous increasing function. The <u>characteristic function</u> T_f is of class C^1 and increases in r and decreases in s. Obviously $\frac{d}{dr} T_f = A_f$.

<u>Lemma 10.4</u>. Assume (B1) - (B4). Let u be a change of scale for τ. Then

$$A_f(r) = (u'(r))^{1-m} \int_{G_r} f^*(\ddot{\omega}) \wedge \varphi_u^{m-1}$$

$$T_f(r,s) = \int_{G_r} \psi_{u,r,s}\, f^*(\ddot{\omega}_0) \wedge \varphi_u^{m-1}$$

for $0 \leqq s < r$. Especially, $T_f(r,s)$ is the characteristic function of f for the bump $(G_r, G_s, \psi_{u,r,s})$ with $\chi = \varphi_u^{m-1}$ if $s \in E_\tau$ and $r \in E_\tau$ and $0 < s < r < +\infty$ resp. $s = 0$ and $r \in E_\tau$.

Proof. Let F be the graph of the meromorphic map $f: M \to \mathbb{P}(V)$. Let $\hat{f}: F \to P(V)$ and $\tilde{f}: F \to M$ projections. Let E'_τ be the set of all r such that r is a regular value of

$\tau \circ \tilde{f} \mid \mathfrak{R}(F)$ and such that $(\tau \circ \tilde{f})^{-1}(r) = \tilde{\Gamma}_r$ has locally finite $\mathfrak{H}^{2m-1}$-measure. Define $\tilde{G}_r = \tilde{f}^{-1}(G_r)$. Then $d\tilde{G}_r = \tilde{\Gamma}_r \cap \mathfrak{R}(F)$. Hence $\tilde{G}_r$ is a Stokes domain on F. Therefore

$$\int_{G_r} f^*(\ddot{\omega}_0) \wedge \varphi_u^{m-1} = \int_{\tilde{G}_r} \tilde{f}^* f^*(\ddot{\omega}_0) \wedge \tilde{f}^*(\varphi_u^{m-1}) =$$

$$= \int_{\tilde{G}_r} \hat{f}^*(\ddot{\omega}_0) \wedge \tilde{f}^*(\varphi_u^{m-1}) =$$

$$= \frac{1}{2\pi} \int_{\tilde{G}_r} d(\hat{f}^*(\ddot{\omega}_0) \wedge (u' \circ \tau \circ \tilde{f})^{m-1} d^c\tau \wedge \varphi^{m-2})$$

$$= (u'(r))^{m-1} \frac{1}{2\pi} \int_{\tilde{\Gamma}_r} \hat{f}^*(\ddot{\omega}_0)\, d^c\tau \wedge \varphi^{m-2}$$

$$= (u'(r))^{m-1} \frac{1}{2\pi} \int_{G_r} f^*(\ddot{\omega}_0) \wedge \varphi^{m-1} = (u'(r))^{m-1} A_f(r)$$

for almost all r, hence for all r. Take $0 \leqq s < r$. Then

$$\int_{G_r} \psi_{u,r,s}\, f^*(\ddot{\omega}_0) \wedge \varphi_u =$$

$$= \int_{G_r - G_s} (\eta_u(r) - \eta_u \circ \tau) f^*(\ddot{\omega}_0) \wedge \varphi_u^{m-1} + \int_{G_s} (\eta_u(r) - \eta_u(s)) f^*(\ddot{\omega}_0) \wedge \varphi_u^{m-1}$$

$$= \int_{G_r - G_r} \int_{\tau(x)}^{r} \eta_u'(t)dt\; f^*(\ddot{\omega}_0) \wedge \varphi_u^{m-1} + \int_{G_s} \int_{s}^{r} \eta_u'(t)dt\; f^*(\ddot{\omega}_0) \wedge \varphi_u^{m-1}$$

$$= \int_s^r \int_{G_t - G_r} f^*(\ddot{\omega}_0) \wedge \varphi_u^{m-1}\, \eta_u'(t)dt + \int_s^r \int_{G_s} f^*(\ddot{\omega}_0) \wedge \varphi_u^{m-1}\, \eta_u'(t)dt$$

$$= \int_s^r \left(\int_{G_t} f^*(\ddot{\omega}_0) \wedge \varphi_u^{m-1}\right) (u'(t))^{m-1} dt$$

$$= \int_s^r A_f(t)\, dt$$

q.e.d.

Let ν be a divisor on M. Define the counting function of ν by

$$n_\nu(r) = \sum_{z \in G_r} \nu(z) \qquad \text{for } r \geqq 0 \text{ if } m = 1$$

$$n_\nu(r) = \int_{\gamma(\nu) \cap G_r} \nu\, \varphi^{m-1} \qquad \text{for } r \geqq 0 \text{ if } m > 1$$

$$N_\nu(r,s) = \int_s^r n_\nu(t)\, dt \qquad \text{for } 0 \leqq s < r$$

$$N_\nu(r) = N_\nu(r,0) = \int_0^r n_\nu(t)\, dt \qquad \text{for } r \geqq 0$$

The counting function n_ν is left semi-continuous. If $\nu \geqq 0$, then $n_\nu \geqq 0$ increases. The valence function is continuous. If $\nu \geqq 0$, then $N_\nu \geqq 0$ increases in r and decreases in r. Both n_ν and N_ν are additive in the divisor ν.

Lemma 10.5. Assume (B1) - (B4). Let ν be a divisor on M. Let u be a change of scale for τ. If $m > 1$ and $r > 0$, then

$$n_\nu(r) = (u'(r))^{1-m} \int_{\gamma(\nu) \cap G_r} \nu\, \varphi_u^{m-1}$$

If $0 \leqq s < r$, then

$$N_\nu(r,s) = \int_{\gamma(\nu) \cap G_r} \psi_{urs}\, \nu\, \varphi_u^{m-1} = \int_{\gamma(\nu) \cap G_r} \psi_{rs}\, \nu\, \varphi^{m-1}$$

Proof. Let $E_\tau(\nu)$ be the set of all $r \geqq 0$ such that $\Gamma_r \cap \gamma(\nu)$ has locally finite $\mathfrak{H}^{2m-3}$-measure and such that r is a regular value of $\tau | \mathfrak{R}(\gamma(\nu))$. Then $\mathfrak{R}(\gamma(\nu)) \cap \Gamma_r = \gamma_r$

is a boundary manifold of $G_r \cap \mathfrak{R}(\gamma(\nu))$. Hence $G_r(\nu) = G_r \cap \gamma(\nu)$ is a Stokes domain if $r \in E_\tau(\nu)$. Take $r \in E_\tau(\nu)$, then

$$\begin{aligned}\int_{G_r(\nu)} \nu\, \varphi_u^{m-1} &= \int_{G_r(\nu)} d\left(\nu \frac{1}{2\pi} (u' \circ \tau)^{m-1} d^c\tau \wedge \varphi^{m-1}\right)\\ &= (u'(r))^{m-1} \int_{\gamma_r} \nu\, d^c\tau \wedge \varphi^{m-1}\\ &= (u'(r))^{m-1} \int_{G_r(\nu)} \nu\, \varphi^{m-1}\\ &= (u'(r))^{m-1}\, n_\nu(r)\end{aligned}$$

if $m > 1$. Take any $r > 0$. A sequence $\{r_\lambda\}_{\lambda \in \mathbb{N}}$ with $r_\lambda \in E_\tau(\nu)$ and $r_\lambda < r_{\lambda+1} < r$ exists such that $r_\lambda \to r$ for $\lambda \to +\infty$. Then $G_r = \bigcup_{\lambda \in \mathbb{N}} G_{r_\lambda}$ and $G_{r_\lambda} \subset G_{r_{\lambda+1}}$. Hence

$$\int_{G_{r_\lambda}(\nu)} \nu\, \varphi_u^{m-1} \to \int_{G_r(\nu)} \nu\, \varphi_u^{m-1} \qquad \text{for } r \to \infty$$

$$n_\nu(r_\lambda) \to n_\nu(r) \qquad \text{for } r \to \infty$$

Therefore

$$\int_{G_r(\nu)} \nu\, \varphi_u^{m-1} = (u'(r))^{m-1}\, n_\nu(r)$$

for all $r > 0$. If $0 \leqq s < r$, then

$$\int_{G_r(\nu)} \psi_{urs}\, \nu\, \varphi_u^{m-1} = \int_{G_r(\nu)-G_s(\nu)} \int_{\tau(x)}^{r} \eta_u'(t)\,dt\; \nu\, \varphi_u^{m-1}$$

$$+ \int_{G_s(\nu)} \int_s^r \eta'_u \, (t)dt\, \nu\, \varphi_u^{m-1}$$

$$= \int_s^r \Big(\int_{G_t(\nu)-G_s(\nu)} \nu\, \varphi_u^m \Big)\, \eta'_u(t)dt + \int_s^r \int_{G_s(\nu)} \nu\, \varphi_u^{m-1}\, \eta'_u(t)\, dt$$

$$= \int_s^r \Big(\int_{G_t(\nu)} \nu\, \varphi_u^m \Big)(u'(r))^{m-1} dt = \int_s^r n_\nu(t)dt = N_\nu(r,s)$$

q.e.d.

Take $a \in \mathbb{P}(V^*)$. Assume that f is adaptable to a. Then, the non-negative divisor $\nu = \delta_f^a$ is defined on M. Define

$$n_f(r;a) = n_\nu(r)$$

$$N_f(r,s;a) = N_\nu(r,s) \qquad N_f(r;a) = N_\nu(r)$$

According to Lemma 10.5, $N_f(r,s,a)$ is the valence function for the bump (G_r, G_s, ψ_{urs}) with $\chi = \varphi_u^{m-1}$ and $r \in E_\tau$ and $s \in E_\tau$ and $0 \leqq s < r$.

Take $r \in E_\tau$. Let $j_r\colon \Gamma_r \to M$ be the inclusion. Then

$$j_r^*(d^\perp \psi_{urs} \wedge \varphi_u^{m-1}) = (u'(r))^{1-m}\, j_r^*(d^c\tau \wedge \varphi_u^{m-1})$$

$$= (u'(r))^{-m}\, j_r^*(d^c(u \circ \tau) \wedge \varphi_u^{m-1}$$

$$= j_r^*(d^c\tau \wedge \varphi^{m-1})$$

and

$$j_s^*(d^\perp \psi_{urs} \wedge \varphi_u^{m-1}) = (u'(s))^{-m}\, j_r^*(d^c(u \circ \tau) \wedge \varphi_u^{m-1}) =$$

$$= j_r^*(d^c\tau \wedge \varphi^{m-1})$$

Define the compensation function $m_f(r,a)$ for $r \in E_\tau$ and $a \in \mathfrak{A}_f \subseteq \mathbb{P}(V^*)$ by

$$m_f(r,a) = \frac{1}{2\pi} \int_{\Gamma_r} \log \frac{1}{||f;a||} \; d^c\tau \wedge \varphi^{m-1}$$

If u is a change of scale for τ, then

$$m_f(r,a) = (u'(r))^{-m} \frac{1}{2\pi} \int_{\Gamma_r} \log \frac{1}{||f;a||} \; d^c(u \circ \tau) \wedge \varphi_u^{m-1}$$

If $r \in E_\tau$ and $s \in E_\tau$ with $0 < s < r$, then $m_f(r,a) = m_f(dG_r;a)$ and $m_f(s;a) = m_f^0(dG_s;a)$ are the compensation functions for the bump (G_r, G_s, ψ_{urs}) and for $\chi = \varphi_u^{m-1}$. Observe $m_f^0(dG_0;a) = 0 = m_f(0;a)$.

Take $a \in \mathbb{P}(V^*)$. Assume that f is adapted to a, i.e. $a \in \mathfrak{A}_f$. The deficit is defined by

$$D_f(r;a) = \int_{G_r} \log \frac{1}{||f;a||} \; \varphi^m \geqq 0$$

for $r \geqq 0$.

Lemma 10.6. Assume (B1) - (B4). Let u be a change of scale. Take $0 \leqq s < r$. Then

$$\frac{1}{2\pi} dd^\perp \psi_{urs} \wedge \varphi_u^{m-1} = \varphi^m \qquad \text{on } G_r - \overline{G}_s$$

If f is adapted to $a \in \mathbb{P}(V^*)$, then

$$D_f(r,a) - D_f(s,a) = \frac{1}{2\pi} \int_{G_r - G_s} \log \frac{1}{||f;a||} \; dd^\perp \psi_{urs} \wedge \varphi_u^{m-1}$$

Especially, $D_f(r,a) - D_f(s,a)$ is the deficit function for the bump (G_r, G_s, ψ_{urs}) with $\chi = \varphi_u^{m-1}$.

Proof. On $G_r - \overline{G}_s$

$$\frac{1}{2\pi} dd^{\perp} \psi_{urs} \wedge \varphi_u^{m-1} = \frac{1}{2\pi} d(\eta_u' \circ \tau \, d^{\perp}\tau \wedge \varphi_u^{m-1})$$

$$= \frac{1}{2\pi} d(\eta_u' \circ \tau)(u'(t))^{m-1} d^{\perp}\tau \wedge \varphi^{m-1})$$

$$= \frac{1}{2\pi} dd^{\perp}\tau \wedge \varphi^{m-1} = \varphi \wedge \varphi^{m-1} = \varphi^m$$

which implies the reminder of the Lemma immediately;

q.e.d.

Therefore, the First Main Theorem reads

$$T_f(r,s) = N_f(r,s,a) + m_f(r,a) - m_f(s,a) - D_f(r,a) + D_f(s,a)$$

It suffices to state the theorem formally if $s = 0$.

<u>Theorem 10.7</u>. <u>First Main Theorem</u>.
Assume (B1) - (B4). Take $r \in E_\tau$. Let f be adapted to $a \in \mathbb{P}(V^*)$. Then

$$T_f(r) = N_f(r,a) + m_f(r,a) - D_f(r,a)$$

The following, additional general assumptions will be made

(B5) Let B be a holomorphic differential form of bidegree $(m-1,0)$ on M. Let $h' = h_\alpha' = D_{B,\alpha}h$ be the associated contravariant differential operator.

(B6) Assume that f is general for B.

A convex change of scale $u \in \mathfrak{U}$ is said to majorize B if and only if for every compact subset K of M, there exists a constant $C > 0$ such that

$$i_0 B \wedge \overline{B} \leqq C \varphi_u^{m-1} \quad \text{on } K.$$

Call $u \in \mathfrak{U}$ a strong convex change of scale if and only if $\varphi_u > 0$ on M. If $u \in \mathfrak{U}$ is a strong convex change of scale then u majorizes B. A convex change of scale u is said to strictly majorize B if and only if

$$i_0 B \wedge \overline{B} \leqq \varphi_u^{m-1}$$

on M.

Now, let u be a convex change of scale then $\hat{G}_r = G_r \cup \Gamma_r$ is compact. If $r > r_0$, then $G_r \neq \emptyset$ and

$$Y_u(r) = \inf \{C \mid C \geqq 1 \text{ and } i_0 B \wedge \overline{B} \leqq (C \varphi_u)^{m-1} \text{ on } \hat{G}_r\}$$

exists with $Y_u(r) \geqq 1$. If $z \in M$, then $\tau(z) = r$ and $z \in \hat{G}_r$. Hence

$$i_0 B \wedge \overline{B} \leqq (Y_u \circ \tau \, \varphi_u)^{m-1} \quad \text{on } M.$$

If $G_r = \emptyset$, define $Y_u(r) = 1$. Obviously, Y_u is an increasing function.

If $0 \leqq c \in \mathbb{R}$, let $\mathfrak{U}_c$ be the set of all increasing, unbounded function v of class C^∞ on $\mathbb{R}[c,+\infty)$ such that $v(x) \geqq 0$, and $v'(x) > 0$ and $v''(x) \geqq 0$ for all $x \geqq c$. Obviously $\mathfrak{V} = \mathfrak{V}_0$. If $u_1 \in \mathfrak{V}$ and $u_0 \in \mathfrak{V}$ then $u_0 \prec u_1$ if and only if $v \in \mathfrak{V}_c$ with $c = u_0(0)$ exists such that $u_1 = v \circ u_0$. This defines a partial order on $\mathfrak{V}$. Observe that u_1 and u_0 are increasing. Hence v is unique.

<u>Lemma 10.8</u>. Assume (B1) and (B2). Take $u \in \mathfrak{V}$ and $v \in \mathfrak{V}_c$ with $c = u(0)$. Then $v \circ u \in \mathfrak{V}$ and

$$\varphi_{v \circ u} \geqq (v' \circ u \circ \tau)\, \varphi_u$$

Proof. Obviously, $v \circ u \in \mathfrak{V}$. Also

$$\varphi_{v \circ u} = \frac{1}{2\pi} dd^c\, v \circ u \circ \tau = (v' \circ u \circ \tau)\, \varphi_u + \frac{1}{2\pi} (v'' \circ u \circ \tau)\, du \circ \tau \wedge d^c u \circ \tau$$

$$\geqq (v' \circ u \circ \tau)\, \varphi_u$$

q.e.d.

<u>Lemma 10.9</u>. Assume (B1), (B2) and (B5). Take $u_0 \in \mathfrak{V}$ and $u \in \mathfrak{V}$ with $u < u_0$. Assume $u'(r) Y_u(r) \leqq u_0'(r)$ for $r \geqq r_0$. Then u_0 strictly majorizes B.

Proof. Define $c = u(0)$. Then $v \in \mathfrak{V}_c$ exists uniquely such that $u_0 = v \circ u$. Then $u_0' = (v' \circ u)\, u'$. Hence

$$i_0\, B \wedge \overline{B} \leqq (Y_u \circ \tau\, \varphi_u)^{m-1} \leqq \left(\frac{u_0' \circ \tau}{u' \circ \tau}\, \varphi_u\right)^{m-1} = ((v' \circ u)\, \varphi_u)^{m-1}$$

$$\leqq (\varphi_{v \circ u})^{m-1} = \varphi_{u_0}^{m-1}$$

q.e.d.

Proposition 10.10. Assume (B1), (B2) and (B5). Let u be a convex change of scale which majorizes B. Take $\varepsilon > 0$. Then there exists a convex change of scale u_0 with $u < u_0$, which strictly majorizes B, such that

$$Y_u(r+\varepsilon) + \varepsilon > \frac{u_0'(r)}{u'(r)} \geqq Y_u(r) \qquad \text{for all } r \geqq 0.$$

Proof. An increasing function $h \geqq 0$ on $\mathbb{R}[0,+\infty)$ is defined by

$$h(r) = \int_0^1 Y_u(r + \varepsilon t)\, dt$$

Then

$$1 \leqq Y_u(r) \leqq h(r) \leqq Y_u(r+\varepsilon) \quad \text{for all } r \geqq 0$$

Also

$$h(r) = \frac{1}{\varepsilon} \int_r^{r+\varepsilon} Y(x)\, dx.$$

Hence h is continuous. Take an increasing, unbounded sequence $\{r_\nu\}_{\nu\in\mathbb{N}}$ with $r_1 = 0$ and $r_\nu < r_{\nu+1}$ such that

$$h(r_{\nu+1}) \leqq h(r_\nu) + \frac{\varepsilon}{2}$$

for all $\nu \in \mathbb{N}$. Define $\rho_\nu = \frac{1}{3}(r_{\nu+1} - r_\nu)$. Take an increasing function g_ν of class C^∞ on $\mathbb{R}[r_\nu, r_{\nu+1}]$ such that

$$1 \leqq g_\nu(x) = h(r_\nu) + \frac{\varepsilon}{2} \qquad \text{for } r_\nu \leqq x \leqq r_\nu + \rho_\nu$$

$$g_\nu(x) = h(r_{\nu+1}) + \frac{\varepsilon}{2} \qquad \text{for } r_{\nu+1}-\rho_\nu \leqq x \leqq r_{\nu+1}$$

Then one and only one function g of class C^∞ on $\mathbb{R}[0,+\infty)$ exists such that $g|\mathbb{R}[r_\nu, r_{\nu+1}] = g_\nu$. Define $c = u(0) \geqq 0$. Then

$$u: \mathbb{R}[0,+\infty) \to \mathbb{R}[c,+\infty)$$

is a diffeomorphism of class C^∞. Define $\hat{u} = u^{-1}$ as the inverse map. Define a function v of class C^∞ on $\mathbb{R}[c,+\infty)$ by

$$v(r) = \int_c^r (g \circ \hat{u})dx \geqq 0$$

Then $v'(r) = g \circ \hat{u} \geqq 1$ and $v''(r) = (g' \circ \hat{u})\, \hat{u}' \geqq 0$. Also $v(r) \geqq r-c$. Hence $v(r) \to \infty$ for $r \to \infty$. Therefore $v \in \mathfrak{A}_c$ and $u_0 = v \circ u \in \mathfrak{A}$ with $u < u_0$. Also $u_0' = (v' \circ u)\, u'$.

Take $r > 0$. Then $r_\nu \leqq r < r_{\nu+1}$ for exactly one $\nu \in \mathbb{N}$. Also

$$Y_u(r) \leqq h(r) \leqq h(r_{\nu+1}) \leqq h(r_\nu) + \frac{\varepsilon}{2} \leqq g(r) = v'(u(r)) = \frac{u_0'(r)}{u'(r)}$$

$$\frac{u_0'(r)}{u'(r)} = v'(u(r)) = g(r) \leqq g(r_{\nu+1}) = h(r_{\nu+1}) + \frac{\varepsilon}{2} \leqq h(r_\nu) + \varepsilon$$

$$\leqq h(r + \varepsilon) + \varepsilon \leqq Y(r + \varepsilon) + \varepsilon$$

q.e.d.

If f_p is adjusted to $a \in \mathbb{P}(V^*)$, define

$$J_p(r;a) = \int_{G_r} \frac{||f_{p+1};a||^2}{||f_p;a||^{2\lambda}} H_p$$

for all $r \geqq 0$ and $0 \leqq p < n$ and $0 < \lambda < 1$. Then $J_p(r,a)$ is continuous and increasing in r. Also the following general assumption shall be made.

(B7) A convex change of scale u is given, which majorizes B. Define $Z_u = Y_u \cdot u' > 0$.

Obviously, Z_u is a positive increasing function on $R(0,+\infty)$.

Theorem 10.11. Assume (B1) - (B7). Take $0 < \lambda < 1$. Let p be an integer with $0 \leqq p < n$. Take r and s with $0 \leqq s < r$. Then

$$\int_s^r J_p(t,a) \frac{dt}{(Z_u(t))^{m-1}} \leqq \frac{c_n\, e^p}{(1-\lambda)^2} [T_{f_p}(r,s) + \Phi(r)]$$

where $c_n > 2$ depends on n only and increases with n.

Proof. By continuity, it suffices to prove the statement for $r \in E_\tau$ and $s \in E_\tau$ with $0 \leqq s < r$. Take any $\varepsilon > 0$. By Proposition 10.10 a convex change of scale $u_0 \in \mathfrak{U}$ exists such that $u \prec u_0$ and such that

$$Y(r) \leqq \frac{u_0'(r)}{u'(r)} \leqq Y(r+\varepsilon) + \varepsilon$$

Moreover, u_0 strictly majorizes B. The assumptions (A1) - (A7) are satisfied for the bump $(G_r, G_s, \psi_{u_0 rs})$ with $\chi = \varphi_{u_0}^{m-1}$. By Lemma 10.6

$$\frac{1}{2\pi}\, dd^{\perp}\, \psi_{u_0 rs} \wedge \varphi_{u_0}^{m-1} = \varphi^m \geqq 0$$

on $G_r - \overline{G}_s$. Therefore

$$\Phi(G_r) = \Phi^+(G_r) = \frac{1}{4\pi} \int_{G_r - G_s} dd^{\perp}\psi_{u_0 rs} \wedge \varphi_{u_0}^{m-1} + \frac{1}{4\pi} \int_{\Gamma_s} d^{\perp}\psi_{u_0 rs} \wedge \chi$$

$$= \frac{1}{2} \int_{G_r - G_s} \varphi^m + \frac{1}{4\pi} \int_{\Gamma_s} d^c(\eta_{u_0} \circ \tau) \wedge \varphi_{u_0}^{m-1}$$

$$= \frac{1}{2} \int_{G_r - G_s} \varphi^m + \eta'_{u_0}(r)\, (u_0'(r))^{m-1} \frac{1}{4\pi} \int_{\Gamma_s} d^c\tau \wedge \varphi^{m-1}$$

$$= \frac{1}{2} \int_{G_r - G_s} \varphi^m + \frac{1}{4\pi} \int_{G_s} (dd^c\tau) \wedge \varphi^{m-1}$$

$$= \frac{1}{2} \int_{G_r} \varphi^m = \frac{1}{2}\, \Phi(r) \leqq \Phi(r)$$

Theorem 9.16 implies

$$\int_{G_r} \psi_{u_0 rs} \frac{||f_{p+1};a||^2}{||f_p;a||^{2\lambda}}\, H_p \leqq \frac{c_n e^p}{(1-\lambda)^2} \; [T_{f_p}(r,s) + \Phi(r)]$$

Here

$$\int_{G_r} \psi_{u_0 rs} \frac{||f_{p+1};a||^2}{||f_p,a||^{2\lambda}}\, H_p =$$

$$= \int_{G_r - G_s} \int_{\tau(x)}^{r} \eta'_{u_0}(t)dt \frac{||f_{p+1};a||^2}{||f_p;a||^{2\lambda}}\, H_p + \int_{G_s} \int_s \eta'_{u_p}(t)dt \frac{||f_{p+1};a||^2}{||f_p;a||^{2\lambda}}\, H_p$$

$$= \int_s^r J_p(t;a)\, (u_0'(t))^{1-m}\, dt$$

$$\geqq \int_s^r J_p(t;a)\, [Y_u(t+\varepsilon) + \varepsilon)\, u'(t)]^{1-m}\, dt$$

Because Y_u increases, $Y_u(t+\varepsilon)+\varepsilon \to Y_u(t)$ for $\varepsilon \to 0$ monotonically for almost all $t \in \mathbb{R}[s,r]$. Therefore $\varepsilon \to 0$ implies

$$\int_s^r J_p(t;a)\ [Y_u(t)u'(t)]^{1-m}\,dt \leqq \frac{c_n e^p}{(1-\lambda)^2}\,[T_{f_p}(r,s) + \Phi(r)]$$

q.e.d.

XI. FIRST DEFICIT ESTIMATES

Several types of deficit estimates will be given. The first type estimates the deficit by the characteristic functions T_{f_p} and by other invariants. The second type of estimates replaces T_{f_p} by T_f via the Plückert difference formula. The first type will be proved on pseudo-convex spaces, where upon the second type will be proved on Stein manifolds only. These types of estimates are subdivided into two classes, uniform estimates and asymptotic estimates. In the first case $D_f(r;a)$ is estimated by invariants evaluated at θr where $\theta > 1$ is some constant. The second case gives estimates evaluated at r, but only outside an exceptional set of finite measure. Uniform estimates hold uniformly in a and the constants do not depend on f. For rapid growth, uniform estimates are less sharp than asymptotic estimates. Uniform estimates are not of much help for defect estimates, but asymptotic estimates are disadvantaged by the fact that the exceptional sets depend uncontrollably on a.

<u>Theorem 11.1</u>. Assume (B1) - (B7). Take $0 < \lambda < 1 < \theta$. Let p be an integer with $0 \leqq p < n$. Take $a \in \mathbb{P}(V^*)$. Take

Take $r > 0$. Then

$$J_p(r;a) \leqq \frac{c_n e^p}{(1-\lambda)^2} \frac{(Z_u(\theta r))^{m-1}}{r(\theta-1)} [T_{f_p}(\theta r) + \Phi(\theta r)]$$

where c_n depends on n only and increases with n.

Proof.

$$r(\theta-1)\, J_p(r,a) \leqq [Z_u(\theta r)]^{m-1} \int_r^{r\theta} J_p(t,a)[Z_u(t)]^{1-m}dt$$

$$\leqq \frac{c_n e^p}{(1-\lambda)^2} (Z_u(\theta r))^{m-1} [T_{f_p}(\theta r) + \Phi(\theta r)]$$

q.e.d.

Let g and h be functions on $\mathbb{R}[a,+\infty]$. Then $g = h$ respectively $g \leqq h$ hold <u>except for a set of finite measure (E.F.M.)</u> if and only if there exists a measurable subset E of $\mathbb{R}[0,+\infty)$ with finite Lebesgue measure such that $g(x) = h(x)$ respectively $g(x) \leqq h(x)$ for all $x \in R[a,+\infty) - E$. Write

$$g \mathrel{\underset{\bullet}{=}} h \qquad\qquad \text{resp.} \qquad g \mathrel{\underset{\bullet}{\leqq}} h$$

<u>Lemma 11.2</u>. Let $a \in \mathbb{R}$. Let h be a non-negative function on $\mathbb{R}[a,+\infty)$ which is locally integrable. Define H by $H(x) = \int_a^x h(t)dt \geqq 0$. Then $h \mathrel{\underset{\bullet}{\leqq}} H^2$.

Proof. The well known Lemma is proved here for the reader's convenience. If $H \equiv 0$, the statement is trivial. Assume $H \not\equiv 0$. Then $s > a$ exists such that $H(s) \neq 0$. Then $H(x) \neq 0$ for all $x \geqq s$. Let E_1 be the set of all $x \geqq a$ such that either $H'(x)$ does not exist or such that $H'(x) \neq h(x)$.

The set E_1 has measure zero. Let E_2 be the set of all $x \geqq a$ with $x \notin E_1$ such that $H(x)^2 \leqq h(x)$. Define $E = E_1 \cup E_2 \cup \mathbb{R}[a,s]$. Then E is measurable. Define $E[r] = E \cap \mathbb{R}[0,r]$. If $r > s$, then

$$\int_{E[r]} dx \leqq s-a + \int_{E[r]-E[s]} \frac{h(x)}{H(x)^2}\, dx \leqq (s-a) + \int_{E[r]-E[s]} \frac{H(x)}{H(x)^2} dx$$

$$\leqq (s-a) + \int_{H(s)}^{\infty} \frac{dy}{y^2} = s-a + \frac{1}{H(s)}$$

Therefore $\int_E dx < +\infty$. If $x \in \mathbb{R}[a,+\infty)-E$, then $h(x) < H(x)^2$;

q.e.d.

Theorem 11.3. Assume (B1) - (B7). Take $0 < \lambda < 1$. Let p be an integer with $0 \leqq p < n$. Take $a \in \mathbb{P}(V^*)$. Then

$$J_p(r,a) \leqq \frac{c_n^2\, e^{2p}}{(1-\lambda)^4} (Z_u(r))^{m-1} [T_{f_p}(r) + \Phi(r)]^2$$

except for a set of finite measure on $\mathbb{R}[0,+\infty)$.

Proof. According to Lemma 11.4 and Theorem 10.11

$$J_p(r,a)(Z_u(r))^{1-m} \leqq \left(\int_0^r J_p(t,u)(Z_u(t))^{1-m} dt\right)^2$$

$$\leqq \frac{c_n^2\, e^{2p}}{(1-\lambda)^4} (T_{f_p}(r) + \Phi(r))^2$$

for all $r \in \mathbb{R}[0,+\infty)-E$ where $\int_E dx < +\infty$;

q.e.d.

Assume (B1) - (B6). Let u be a change of scale for τ. If ξ is a real differential form of bidegree (q,q) on M, define

$$\text{pos}(\xi) = \{x \in M \mid \xi(x) > 0\}$$

Obviously $\text{pos}(\varphi_u^m)$ is open and $M-\text{pos}(\varphi_u^m)$ has measure zero. A non-negative function h_{up} of class C^∞ is defined on $\text{pos}(\varphi_u^m) - I_{f_p}$ by

$$H_p = h_{up}\,\varphi_u^m$$

Similarly, $h_p = h_{id,p}$ is defined on $\text{pos}(\varphi^m) - I_{f_p}$ by

$$H_p = h_p\,\varphi^m$$

Lemma 11.4. Assume (B1) - (B6). Let u be a change of scale for τ. Then $(\log h_{up})\,\varphi_u^m$ is locally integrable on M.

Proof. If $x \geqq 0$, then $\log^+ x \leqq 2x^{\frac{1}{2}}$. Hence

$$|\log x| = \log^+ x + \log^+ \frac{1}{x} \leqq 2(x^{\frac{1}{2}} + x^{-\frac{1}{2}}) = 2x^{-\frac{1}{2}}(1+x)$$

for $x > 0$. Let $\alpha \in \mathfrak{T}_M$ be a patch such that there exists an irreducible representation $\mathfrak{w} : U_\alpha \to V$ of f. Define $\mathfrak{w}_q = \mathfrak{w}_{q\alpha}$ and

$$\upsilon_\alpha = \left(\frac{i}{2\pi}\right)^m d\alpha_1 \wedge d\bar{\alpha}_1 \wedge \ldots \wedge d\alpha_m \wedge d\bar{\alpha}_m$$

Then $H_p = H_p^\alpha\,\upsilon_\alpha$ and $H_p^\alpha = |\mathfrak{w}_{p-1}|^2\,|\mathfrak{w}_{p+1}|^2\,|\mathfrak{w}_p|^{-4}$. Therefore $\log |H_p^\alpha|\,\upsilon_\alpha$ is locally integrable over U_α. Also $\varphi_u^m = \varphi_\alpha\,\upsilon_\alpha$ where $\varphi_\alpha \geqq 0$ is a function of class C^∞ on U_α and where $\varphi_\alpha > 0$

almost everywhere on U_α. Here $h_{up} = H_p^\alpha \varphi_\alpha^{-1}$ almost everywhere on U_α. Because

$$|\log \varphi_\alpha| \varphi_\alpha \leqq 2 \varphi_\alpha^{\frac{1}{2}} (1 + \varphi_\alpha)$$

the form $(\log \varphi_\alpha) \varphi_\alpha \upsilon_\alpha$ and $(\log H_p^\alpha) \varphi_\alpha \upsilon_\alpha$ are locally integrable over U_α. Hence $\log h_{up} \varphi_u^m = (\log H_p^\alpha - \log \varphi_\alpha) \varphi_\alpha \upsilon_\alpha$ is locally integrable over U_α and those U_α over M. Therefore $(\log h_{up}) \varphi_u^m$ is locally integrable cover M; q.e.d.

The pth inflections of f are defined by

$$S_f^p(r) = \frac{1}{2} \int_{G_r} \log \frac{1}{h_p} \varphi^m$$

$$\overset{+}{S}{}_f^p(r) = \frac{1}{2} \int_{G_r} \log^+ \frac{1}{h_p} \varphi^m \geqq 0$$

If u is a chance of scale for τ, define

$$S_f^p(r;u) = \frac{1}{2} \int_{G_r} \log \frac{1}{h_{up}} \varphi^m$$

$$\overset{+}{S}{}_f^p(r;u) = \frac{1}{2} \int_{G_r} \log^+ \frac{1}{h_{up}} \varphi^m \geqq 0$$

Theorem 11.5. First Uniform Deficit Estimate.
Assume (B1) - (B7). Then there exists a constant $c > 0$ depending on n only, such that for all $\theta > 1$, all $r > r_0$ and all $a \in \mathbb{P}(V^*)$ the following estimate holds

$$D_f(r,a) \leqq 2^n \Phi(r) \sum_{p=0}^{n-1} \log^+ \frac{T_{f_r}(\theta r) + \Phi(\theta r)}{r(\theta-1) \ \Phi(r)} +$$

$$+ 2^n \sum_{p=0}^{n-1} \overset{+}{S}{}_f^p(r) + \Phi(r)\,[c + n(m-1)2^{n-1} \log^+ Z_u(\theta r)]$$

Proof.[5] Recal $||f_n;a|| = 1$. Take $0 < \lambda < 1$. Then

$$\log \frac{1}{||f;a||^2} = \sum_{p=0}^{n-1} \frac{1}{\lambda^{p+1}} \log \frac{||f_{p+1};a||^2}{||f_p;a||^{2\lambda}}$$

Therefore

$$D_f(r;a) = \frac{1}{2} \int_{G_r} \log \frac{1}{||f;a||^2} \quad \varphi^m$$

$$= \frac{1}{2} \sum_{p=0}^{n-1} \frac{1}{\lambda^{p+1}} \int_{G_r} \log \left(\frac{||f_{p+1};a||^2}{||f_p;a||^{2\lambda}} h_p\right) \varphi^m$$

$$+ \frac{1}{2} \sum_{p=0}^{n-1} \frac{1}{\lambda^{p+1}} \int_{G_r} \log \frac{1}{h_p} \quad \varphi^m$$

The convexity of the logarithm implies

$$D_f(r;a) \leqq \frac{1}{2} \sum_{p=0}^{n-1} \frac{\Phi(r)}{\lambda^{p+1}} \log \left[\frac{1}{\Phi(r)} \int_{G_r} \frac{||f_{p+1};a||^2}{||f_p;a||^{2\lambda}} H_p\right]$$

$$+ \sum_{p=0}^{n-1} \frac{1}{\lambda^{p+1}} \overset{+}{S}{}_f^p(r)$$

$$\leqq \Phi(r) \sum_{p=0}^{n-1} \frac{1}{2\lambda^{p+1}} \log \left[\frac{c_n \; e^p}{(1-\lambda)^2} \frac{(Z_u(\theta r))^{m-1}}{r(\theta-1)\Phi(r)} (T_{f_p}(\theta r) + \Phi(\theta r))\right]$$

$$+ \sum_{p=0}^{n-1} \frac{1}{\lambda^{p+1}} \overset{+}{S}{}_f^p(r)$$

$$\leqq \Phi(r) \sum_{p=0}^{n-1} \frac{1}{2\lambda^{p+1}} \log^+ \frac{T_{f_p}(\theta r) + \Phi(\theta r)}{r(\theta-1)\;\Phi(r)} + \frac{1}{\lambda^n} \sum_{p=0}^{n-1} \overset{+}{S}{}_f^p(r) +$$

5) The idea of this proof is due to Griffiths [16].

$$+ \frac{n}{2\lambda^n} \Phi(r) [n + 2 \log \frac{1}{1-\lambda} + \log c_n + (m-1) \log^+ Z_u(\theta r)]$$

Now the estimate follows immediately with $\lambda = \frac{1}{2}$ and

$$c = n\, 2^{n-1} [n + 2 \log 2 + \log c_n] > 0$$

which depends on n only (and increases with n); q.e.d.

Theorem 11.6. The First Asymptotic Deficit Estimate. Assume (B1) - (B7). Then a constant $c > 0$ exists depending only on n, such that for all $a \in \mathbb{P}(V^*)$ and for all $r > r_0$ except a set of finite measure the following inequality holds.

$$D_f(r;a) \leqq 2^n \Phi(r) \sum_{p=0}^{n-1} \log \left(1 + \frac{T_{f_p}(r)}{\Phi(r)}\right) + 2^n \sum_{p=0}^{n-1} \overset{+}{S}{}_f^p(r)$$

$$+ n\, 2^{n-1} \Phi(r) [c + \log^+ \Phi(r) + (m-1) \log^+ Z_u(r)]$$

Proof.[5] As in the proof of Theorem 11.5, the following estimate is obtained

$$D_f(r,a) \leqq \Phi(r) \sum_{p=0}^{n-1} \frac{1}{2\lambda^{p+1}} \log \frac{J_f^p(r)}{\Phi(r)} + \sum_{p=0}^{n-1} \frac{1}{\lambda^{p+1}} \overset{+}{S}{}_f^p(r)$$

Theorem 11.3 implies

$$D_f(r;a) \leqq \Phi(r) \sum_{p=0}^{n-1} \frac{1}{2\lambda^{p+1}} \log \left[\frac{c_n^2\, e^{2p}}{(1-\lambda)^4} \frac{(Z_u(r))^{m-1}}{\Phi(r)} (T_{f_p}(r)+\Phi(r))^2\right]$$

$$+ \sum_{p=0}^{n-1} \frac{1}{\lambda^{p+1}} \overset{+}{S}{}_f^p(r)$$

$$\leq \Phi(r) \sum_{p=0}^{n-1} \frac{1}{\lambda^{p+1}} \log \left(1 + \frac{T_{f_p}(r)}{\Phi(r)}\right) + \sum_{p=0}^{n-1} \frac{1}{\lambda^{p+1}} \overset{+}{S}{}_f^p(r)$$

$$+ \frac{n}{2\lambda^n} \Phi(r) \left[2n + 2 \log c_n + 4 \log \frac{1}{1-\lambda} + \log^+ \Phi(r)\right.$$

$$\left. + (m-1) \log Z_u(r)\right]$$

except for a set of finite measure. Take $\lambda = \frac{1}{2}$ and define $c = 2n + 2 \log c_n + 4 \log 2$. Then the estimate of Theorem 11.6 follows; q.e.d.

XII. SECOND DEFICIT ESTIMATE ON STEIN MANIFOLDS

The assumption (B7) is strengthened to

(B8) A strongly convex change of scale u of τ is given on M.

Of course, (B8) implies (B7). So it suffices to assume "(B1) - (B6) and (B8)", but it is simpler and shorter to write "Assume (B1) - (B8)" although this is redundant.

Assumption (B8) implies that M is a Stein manifold. If M is a Stein manifold, then an exhaustion τ exists such that $dd^c\tau > 0$ on M, hence the identity is a strongly convex change of scale of τ in this case.

At first the Plücker difference formula shall be formulated. Assume (B1) - (B8). Let K be the canonical bundle on M. Let κ be a hermitian metric along the fibers of K. If $\alpha = (\alpha_1, \ldots, \alpha_m) \in \mathfrak{T}_M$ is a patch on M, then $d\alpha_1 \wedge \ldots \wedge d\alpha_m$

is a holomorphic frame of K over U_α. Therefore

$$\kappa_\alpha = \kappa(d\alpha_1 \wedge \cdots \wedge d\alpha_m) > 0 \tag{12.1}$$

is a function of class C^∞ on U_α. If $\alpha \in \mathfrak{T}_M$ and $\beta \in \mathfrak{T}_M$ with $U_\alpha \cap U_\beta \neq \emptyset$, then

$$\begin{aligned} d\alpha_1 \wedge \cdots \wedge d\alpha_m &= \Delta_{\alpha\beta}\, d\beta_1 \wedge \cdots \wedge d\beta_m \\ \kappa_\alpha &= |\Delta_{\alpha\beta}|^2\, \kappa_\beta \end{aligned} \tag{12.2}$$

on $U_\alpha \cap U_\beta$. Inversely, if a family $\{\kappa_\alpha\}_{\alpha \in \mathfrak{T}_M}$ of functions $\kappa_\alpha > 0$ of class C^∞ on U_α is given such that (12.2) is satisfied on $U_\alpha \cap U_\beta \neq \emptyset$, then one and only one hermitian metric along the fibers of k exists satisfying (12.1). If $\alpha = (\alpha_1, \ldots, \alpha_m) \in \mathfrak{T}_M$, define

$$\upsilon_\alpha = \left(\frac{i}{2\pi}\right)^m d\alpha_1 \wedge d\bar{\alpha}_1 \wedge \cdots \wedge d\alpha_m \wedge d\bar{\alpha}_m.$$

Then $\varphi_u^m = \varphi_\alpha \upsilon_\alpha$ on U_α, where $\varphi_\alpha > 0$ is a function of class C^∞. If $\alpha \in \mathfrak{T}_M$ and $\beta \in \mathfrak{T}_M$ and $U_\alpha \cap U_\beta \neq \emptyset$, then $\varphi_\beta = \varphi_\alpha |\Delta_{\alpha\beta}|^2$ on $U_\alpha \cap U_\beta$. Hence

$$\frac{1}{\varphi_\alpha} = |\Delta_{\alpha\beta}|^2 \frac{1}{\varphi_\beta} \qquad \text{on } U_\alpha \cap U_\beta \neq \emptyset$$

Therefore there exists one and only one hermitian metric κ along the fibers of K such that

$$\kappa_\alpha = \frac{1}{\varphi_\alpha} \qquad \text{on } U_\alpha$$

This κ is called <u>the hermitian metric along the fibers of K defined by φ_u^m</u>. Then

$$\kappa \cdot \varphi_u^m = \kappa_\alpha \, \varphi_\alpha = 1 \qquad \text{on } U_\alpha$$

Hence $\kappa \cdot \varphi_u^m \equiv 1$ on M. Also $H_p = h_{up} \, \varphi_\alpha \upsilon_\alpha$ on U_α. Hence $\kappa \, H_p = h_{up} \, \kappa_\alpha \varphi_\alpha = h_{up}$ on U_α. Therefore

$$\kappa \, H_p = h_{up}$$

on M. Consider the bump (G_r, G_s, ψ_{urs}) with $\chi = \varphi_u^{m-1}$. Then

$$S_f^p(G_r;\kappa) = \frac{1}{4\pi} \int_{G_r - G_s} \log \frac{1}{h_{up}} \, dd^\perp \psi_{urs} \wedge \varphi_u^{m-1}$$

$$= \frac{1}{2} \int_{G_r - G_s} \log \frac{1}{h_{up}} \, \varphi^m = S_f^p(r;u) - S_f^p(s,u)$$

If $r \in E_\tau$ and $s \in E_\tau$ with $s \leqq r$, then

$$\Omega_f^p(dG_r;\kappa) = \frac{1}{4\pi} \int_{dG_r} \log \, (\kappa \, H_p) \, d^\perp \psi_{urs} \wedge \varphi_u^{m-1}$$

$$= \frac{1}{4\pi} \int_{\Gamma_r} \log h_{up} \, d^c \tau \wedge \varphi^{m-1}$$

$$\Omega_f^p(dG_s;\kappa) = \frac{1}{4\pi} \int_{\Gamma_s} \log h_{up} \, d^c \tau \wedge \varphi^{m-1}$$

Define

$$\Omega_f^p(r,u) = \frac{1}{4\pi} \int_{\Gamma_r} \log h_{up}\ d^c\tau \wedge \varphi^{m-1}$$

for all $r \in E_\tau$. Then $\Omega_f^p(dG_r;\kappa) = \Omega_f^p(r,\kappa)$ and $\Omega_f^p(dG_s;\kappa) = \Omega_f^p(s,\kappa)$. Also define

$$\Omega_f^p(r) = \Omega_f^p(r,\mathrm{Id}) = \frac{1}{4\pi} \int_{\Gamma_r} \log h_p\ d^c\tau \wedge \varphi^{m-1}$$

Again let κ be the hermitian metric along the fibers of K defined by φ_u^m. If $\alpha \in \mathfrak{T}_M$, then

$$\mathrm{Ric}\ \kappa = -\frac{1}{4\pi} dd^c \log \kappa_\alpha = \frac{1}{4\pi} dd^c \log \varphi_\alpha$$

which is also called the Ricci form of the volume form φ_u^m. Define

$$\mathrm{Ric}_u(r) = \int_0^r \left(\int_{G_t} \mathrm{Ric}\ \kappa \wedge \varphi^{m-1}\right) dt \quad \text{for } r \geqq 0.$$

If $t \in E_\tau$, then

$$\begin{aligned}
\int_{G_t} \mathrm{Ric}\ \kappa \wedge \varphi_u^{m-1} &= \int_{G_t} d(\mathrm{Ric}\ \kappa \wedge d^c(u\circ\tau) \wedge \varphi_u^{m-2}) \\
&= \int_{G_t} d((u'(t))^{m-1} \mathrm{Ric}\ \kappa \wedge d^c\tau \wedge \varphi^{m-2} \\
&= (u'(t))^{m-1} \int_{\Gamma_t} \mathrm{Ric}\ \kappa \wedge d^c\tau \wedge \varphi^{m-2} \\
&= (u'(t))^{m-1} \int_{G_t} \mathrm{Ric}\ \kappa \wedge \varphi^{m-1}
\end{aligned}$$

By continuity

$$(u'(t))^{1-m} \int_{G_t} \mathrm{Ric}\,\kappa \wedge \varphi_u^{m-1} = \int_{G_t} \mathrm{Ric}\,\kappa \wedge \varphi^{m-1}$$

holds for all $t \geqq 0$. Consider the bump (G_r, G_s, ψ_{urs}) with $\chi = \varphi_u^{m-1}$. Then

$$\mathrm{Ric}_\kappa\,(G_r) = \int_{G_r} \psi_{urs}\,\mathrm{Ric}\,\kappa \wedge \varphi_u^{m-1}$$

$$= \int_s^r \Big(\int_{G_t} \mathrm{Ric}\,\kappa \wedge \varphi_u^{m-1}\Big)\,\eta_u'(t)\,dt$$

$$= \int_s^r \int_{G_t} \mathrm{Ric}\,\kappa \wedge \varphi^{m-1}\,dt = \mathrm{Ric}_\kappa(r) - \mathrm{Ric}_\kappa(s)$$

Now the following additional general assumption shall be made

(B9) Let κ be the hermitian metric along the fibers of the canonical bundle K on M defined by φ_u^m where u is given by (B8).

Let v_p be the pth stationary divisor of f. Let $0 \leqq p < n$. Take $r \in E_\tau$ and $s \in E_\tau$ with $s < r$. Then Plücker's difference formula reads

$$N_{v_p}(r,s) + T_{f_{p-1}}(r,s) - 2T_{f_p}(r,s) + T_{f_{p+1}}(r,s) =$$

$$= \Omega_f^p(r,u) - \Omega_f^p(r,s) + S_f^p(r,u) - S_f^p(s,u) + \mathrm{Ric}_u(r) - \mathrm{Ric}_u(s)$$

It has to be recorded formally only for $s = 0$.

Theorem 12.1. Plücker's Difference Formula.

Assume (B1) - (B9). Let p be an integer with $0 \leqq p < n$.

Take $r \in E_\tau$. Then

$$N_{V_p}(r)+T_{f_{p-1}}(r)-2T_{f_p}(r)+T_{f_{p+1}}(r)= \Omega_f^p(r;u)+S_f^p(r;u)+\mathrm{Ric}_u(r).$$

Assume (B1) - (B9). For $r \geqq 0$ define

$$\Phi_1(r) = \frac{1}{2\pi} \int_{G_r} d\tau \wedge d^c\tau \wedge \varphi^{m-1}$$

Lemma 12.3.

$$\Phi_1(r) = \int_0^r \Phi(t)\, dt$$

Proof. It suffices to prove the identity for each $r \in E_\tau$. Stokes Theorem implies

$$\Phi_1(r) = \frac{1}{2\pi} \int_{G_r} d(r-\tau) \wedge d^c\tau \wedge \varphi^{m-1} =$$

$$= \frac{1}{2\pi} \int_{G_r} (r-\tau)\, dd^c\tau \wedge \varphi^{m-1}$$

$$= \int_{G_r} (r-\tau)\, \varphi^m$$

$$= \int_0^r \int_{G_t} \varphi^m\, dt = \int_0^r \Phi(t)\, dt$$

q.e.d.

Consequently, $\Phi_1(r) > 0$ if $r > r_0$.

Lemma 12.4. Assume (B1) - (B8). Let p be an integer with $0 \leqq p < n$. Take $r \geqq 0$, then

$$\int_{G_r} H_p \leqq (Z_u(r))^{m-1} A_{f_p}(r)$$

Proof.

$$\int_{G_r} H_p = \int_{G_r} i_0 \, B \wedge \overline{B} \wedge f_p^*(\ddot{\omega}_p) \leqq (Y_u(r))^{m-1} \int_{G_r} f_p^*(\ddot{\omega}_p) \wedge \varphi_u^m$$

$$= (Y_u(r) u'(r))^{m-1} A_{f_p}(r) = (Z_u(r))^{m-1} A_{f_p}(r).$$

Lemma 12.5. Assume (B1) - (B8). Let p be an integer with $0 \leqq p < n$. Then $A_{f_p}(r) > 0$ for $r > r_0$ and $T_{f_p}(r) \to \infty$ for $r \to \infty$.

Proof. Take $r > r_0$. Take $\alpha \in \mathfrak{T}_M$ with $U_\alpha \subseteq G_r$ such that an irreducible representation $\mathfrak{w}: U_\alpha \to V$ of f on U_α exists. A thin analytic set S in U_α exists such that $\mathfrak{w}_q(x) = \mathfrak{w}_{q\alpha}(x) \neq 0$ for all $x \in U_\alpha - S$ and all $q = 0, 1, \ldots, n$. Then $H_p = H_p^\alpha \upsilon_\alpha$ on $U_\alpha - I_{f_p}$ and

$$H_p^\alpha = |\mathfrak{w}_{p-1}|^2 \, |\mathfrak{w}_{p+1}|^2 \, |\mathfrak{w}_p|^{-4} > 0 \quad \text{on } U_\alpha - S$$

For $r > r_0$, the set G_r is not empty. Therefore

$$0 < \int_{G_r} H_p \leqq Z_u(r) \, A_{f_p}(r).$$

Therefore $A_{f_p}(r) > 0$ for $r > r_0$. If $r > r_1 > r_0$, then

$$T_{f_p}(r) = \int_0^r A_{f_p}(t)dt \geqq \int_{r_1}^r A_{f_p}(t)dt \geqq (r - r_1) A_{f_p}(r_1) > 0$$

Hence $T_{f_p}(r) \to +\infty$ for $r \to \infty$; q.e.d.

Observe

$$\liminf_{r \to \infty} \frac{T_{f_p}(r)}{r} > 0$$

is a consequence of the proof.

Lemma 12.6. Assume (B1) and (B2). Then $\Phi_1(r) \to \infty$ for $r \to \infty$.

Proof. If $r > r_0$, then $\Phi(r) > 0$. Take $r > r_1 > r_0$. Then

$$\Phi_1(r) = \int_0^r \Phi(t)dt \geqq \int_{r_1}^r \Phi(t)dt = (r-r_1)\Phi(r_1) > 0$$

Hence $\Phi_1(r) \to +\infty$ for $r \to \infty$. q.e.d.

Plücker's difference formula shall be used to eliminate the characteristic functions $T_{f_1}, \ldots, T_{f_{n-1}}$ of the associated maps from the deficit estimates. At first, the asymptotic case shall be studied.

Asymptotic Estimates

Assume (B1) - (B10). Functions

$$h_{up} : M - I_{f_p} \to \mathbb{R}[0,+\infty)$$

$$h_p : \operatorname{pos}(\varphi^m) - I_{f_p} \to \mathbb{R}[0,+\infty)$$

$$\rho_u : M \to \mathbb{R}[0,+\infty)$$

$$\rho : \operatorname{pos}(\varphi^m) \to \mathbb{R}[0,+\infty)$$

$$\delta_u : M \to \mathbb{R}[0,+\infty)$$

of class C^∞ are defined by

$$H_p = h_{up}\,\varphi_u^m = h_p\,\varphi^m$$

$$\varphi^m = \delta_u\,\varphi_u^m$$

$$\frac{1}{2\pi}\,d\tau \wedge d^c\tau \wedge \varphi^{m-1} = \rho_u\,\varphi_u^m = \rho\varphi^m$$

with

$$h_{up} = \delta_u\,h_p \qquad \text{and } \rho_u = \rho\delta_u$$

Define

$$Q_u(r) = \frac{1}{4\pi}\int_{G_r} \log^+ \rho_u\, d\tau \wedge d^c\tau \wedge \varphi^{m-1} \geqq 0$$

for $r \geqq 0$. The function Q_u increases and is continuous. The integral exists because

$$0 \leqq \frac{1}{2\pi}\,|\log \rho_u|\,d\tau \wedge d^c\tau \wedge \varphi^{m-1} \leqq |\log \rho_u|\,\rho_u\,\varphi_u^m$$

where ρ_u is a function of class C^∞ on M. For $r \in E_\tau$, define

$$\hat{\Omega}_f^p(r;u) = \frac{1}{4\pi}\int_{\Gamma_r} \log\,(1+h_{up})\,d^c\tau \wedge \varphi^{m-1} \geqq 0$$

Then

$$\Omega_f^p(r;u) \leqq \hat{\Omega}_f^p(r;u)$$

<u>Lemma 12.7</u>. Assume (B1) - (B9). Let p be an integer with $0 \leqq p < n$. Take $r > r_0$. Then

$$\int_0^r \hat{\Omega}_f^p(t;u)dt \leqq \frac{1}{2}\,\Phi_1(r)\,\log^+ \frac{\Phi(r)u'(r)+A_{f_p}(r)}{\Phi_1(r)}$$

$$+\frac{1}{2}\,(m-1)\,\Phi_1(r)\,\log^+ Z_u(r) + Q_u(r)$$

Proof. According to [43] Lemma AII 3.3 dG_r carries the opposite orientation to the orientation need for fiberintegration over the fibers of τ. Therefore

$$\int_0^r \hat{\Omega}_f^p(t,u)dt$$

$$= -\frac{1}{4\pi}\int_{G_r} \log\,(1+h_{up})\; d^c\tau \wedge \varphi^{m-1} \wedge d\tau$$

$$= \frac{1}{2}\,\Phi_1(r) \int_{G_r} \log \frac{1+h_{up}}{\rho_u}\;\; \frac{d\tau \wedge d^c\tau \wedge \varphi^{m-1}}{2\pi\,\Phi_1(r)} \;+$$

$$+\frac{1}{4\pi}\int_{G_r} \log \rho_u\; d\tau \wedge d^c\tau \wedge \varphi^{m-1}$$

$$\leqq \frac{1}{2}\,\Phi_1(r)\,\log\left(\int_{G_r} \frac{(1+h_{up})}{\rho_u}\;\frac{\rho_u\,\varphi_u^m}{\Phi_1(r)}\right) + Q_u(r)$$

$$= \frac{1}{2}\,\Phi_1(r)\,\log\left[\frac{1}{\Phi_1(r)}\left(\int_{G_r}\varphi_u^m + \int_{G_r} H_p\right)\right] + Q_u(r)$$

$$\leqq \frac{1}{2}\,\Phi_1(r)\,\log\left[\frac{1}{\Phi_1(r)}(u'(r))^m\Phi(r) + (Y_u(r)u'(r))^{m-1}A_{f_p}(r)\right]$$

$$+ Q_u(r)$$

Because $Y_u(r) \geqq 1$, this implies

$$\int_0^r \hat{\Omega}_f^p(t,u)dt \leqq \frac{1}{2}\,\Phi_1(r)\ \log^+ \frac{1}{\Phi_1(r)}\ (u'(r)\Phi(r) + A_{f_p}(r))$$

$$+ \frac{1}{2}\,(m-1)\ \Phi_1(r)\ Z_u(r) + Q_u(r)$$

q.e.d.

Let $a_\mu \geqq 0$ for $\mu = 1,\ldots,q$. The following well-known inequalities will be needed

$$(a_1 + \ldots + a_q)^2 \leqq q(a_1^2 + \ldots + a_q^2)$$

$$\log^+(a_1 + \ldots + a_q) \leqq \log^+ a_1 + \ldots + \log^+ a_q + \log q$$

$$\log^+(a_1 \cdot a_2\ \ldots\ a_q) \leqq \log^+ a_1 + \ldots + \log^+ a_q$$

Lemma 12.8. Assume (B1) - (B9). Let p be an integer with $0 \leqq p < n$. Then for all $r \geqq 0$ except a set of finite measure, the following inequality holds.

$$\Omega_f^p(r;u) \leqq \hat{\Omega}_f^p(r,u)$$

$$\overset{\leqq}{\bullet}\ \frac{1}{2}\,(\Phi_1(r))^2\ T_{f_p}(r) + 2m^2\ (\Phi_1(r)\log^+ Z_u(r))^2 +$$

$$+ 2\ (\Phi(r))^2 + 5(Q_u(r))^2$$

Proof. Because $T_{f_p}(r) = \int_0^r A_{f_p}(t)dt$, Lemma 11.2 implies

$$A_{f_p}(r)\ \overset{\leqq}{\bullet}\ (T_{f_p}(r))^2$$

Lemma 12.5 implies

$$\int_0^r \hat{\Omega}_f^p(t;u)dt \leqq \frac{1}{2}\Phi_1 \log^+ \frac{\Phi}{\Phi_1} + \frac{1}{2}\Phi_1 \log u' + \frac{1}{2}\Phi_1 \log^+ \frac{A_{f_p}}{\Phi_1}$$

$$+ \frac{1}{2}(m-1)\log^+ Z_u + Q_u + \frac{1}{2}\Phi_1 \log 2$$

$$\leqq \frac{1}{2}\Phi + \frac{1}{2}\Phi_1 \log^+ \frac{A_{f_p}}{\Phi_1} + \frac{1}{2} m \log^+ Z_u + Q_u$$

$$+ \frac{1}{2}\Phi_1 \log 2.$$

because $u' \leqq u'Y_u = Z_u$. Therefore

$$\hat{\Omega}_f^p(r;u) \leqq \frac{1}{4}\left(\Phi + \Phi_1 \log^+ \frac{A_{f_p}}{\Phi_1} + \frac{m}{2}\Phi_1 \log^+ Z_u + 4Q_u + \Phi_1 \log 2\right)^2$$

$$\leqq \frac{5}{4}\Phi^2 + \frac{5}{4}\Phi_1^2\left(\log^+ \frac{A_{f_p}}{\Phi_1}\right)^2 + \frac{5}{4} m^2 \Phi_1^2 (\log^+ Z_u)^2$$

$$+ 5\,Q_u^2 + \frac{5}{4}(\log 2)^2 \Phi_1^2$$

$$\leqq\!\!\!\!\cdot\;\; 2\,\Phi^2 + 5\,\Phi_1^2\left(\log^+ \frac{T_{f_p}}{\Phi_1^{\frac{1}{2}}}\right)^2 + 2m^2\Phi_1^2(\log^+ Z_u)^2$$

$$+ 5\,Q_u^2 + 5\,\Phi_1^2$$

Now, $\log^+ x \leqq 2\log^+ x^{\frac{1}{2}} \leqq 2\,x^{\frac{1}{2}}$ for $x \geqq 0$ implies

$$\hat{\Omega}_f^p(r,u) \leqq\!\!\!\!\cdot\;\; 2\,\Phi^2 + 20\,\Phi_1^{\frac{3}{2}} T_{f_p} + 2m^2\Phi_1^2(\log^+ Z_u)^2 + 5Q_u^2 + 5\,\Phi_1^2$$

According to Lemma 12.5 and Lemma 12.6, a number r_1 exists such that $6400 < \Phi_1(r)$ and $20 < T_{f_p}(r)$ for $r > r_1$. Then

$$\hat{\Omega}_f^p(r;u) \leqq 2\,\Phi^2 + \frac{1}{4}\,\Phi_1^2\,T_{f_p} + 2m^2\,\Phi_1^2\,(\log^+ Z_u)^2 + 5Q_u + \frac{1}{4}\Phi_1^2\,T_{f_p}$$

$$\leqq \frac{1}{2}\,\Phi_1^2\,T_{f_p} + 2m^2\,(\Phi_1 \log^+ Z_u)^2 + 5Q_u + 2\,\Phi^2$$

q.e.d.

For $r > 0$, define $P_r = G_r \cap \operatorname{pos}(\operatorname{Ric}\,\kappa \wedge \varphi^{m-1})$. Define

$$\operatorname{Ric}_u^+(r) = \int_0^r \Big(\int_{P_t} \operatorname{Ric}_\kappa \wedge \varphi^{m-1}\Big)dt \geqq \operatorname{Ric}_u(r)$$

Proposition 12.9. Assume (B1) - (B9). Let p be an integer with $0 \leqq p < n$. Then the following estimate holds for all $r > 0$ except a set of finite measure.

$$T_{f_p} \leqq \Phi_1^{2p}\,T_f + \sum_{q=0}^{p-1} \Phi_1^{2p-2q-2}\,S_f^q(r;u) + p\,\Phi_1^{2p-2}\,L_u$$

where

$$L_u(r) = 2\,\Phi^2(r) + 2m^2(\Phi_1(r)\,\log^+ Z_u(r))^2 + 5(Q_u(r))^2 +$$

$$+ \operatorname{Ric}_u^+(r)$$

is independent of p.

Proof. Correct for $p = 0$. Assume $p \geqq 1$. Plucker's difference formula and Lemma 12.8 imply

$$T_{f_{p+1}}(r) \leqq 2\,T_{f_p}(r) + \Omega_f^p(r;u) + S_f^p(r;u) + \operatorname{Ric}_u^+(r)$$

$$\leqq 2\,T_{f_p}(r) + \frac{1}{2}(\Phi_1(r))^2 T_{f_p}(r) + S_f^p(r,u) + L_u(r)$$

A number $r_1 > r_0$ exists such that $4 < (\Phi_1(r))^2$ if $r > r_1$. Hence

$$T_{f_{p+1}} \underset{\bullet}{\leqq} \Phi_1^2 T_{f_p} + S_f^p(r;u) + L_u$$

Hence

$$T_{f_1} \underset{\bullet}{\leqq} \Phi_1^2 T_f + S_f^p(r;u) + L_u$$

which is the desired estimate for $p = 1$. Assume the case for p is proved. Then the case for $p+1$ shall be proven if $p + 1 < n$.

$$\begin{aligned} T_{f_{p+1}} &\underset{\bullet}{\leqq} \Phi_1^2 \left(\Phi_1^{2p} T_f + \sum_{q=0}^{p-1} \Phi_1^{2p-2q-2} S_f^q(r;u) + p\, \Phi_1^{2p-2} L_u\right) \\ &\quad + S_f^p(r;u) + L_u \\ &\underset{\bullet}{\leqq} \Phi_1^{2p+2} T_f + \sum_{q=0}^{p} \Phi_1^{2p-2q} S_f^q(r;u) + (p\, \Phi_1^{2p} + 1) L_u \\ &\underset{\bullet}{\leqq} \Phi_1^{2p+2} T_f + \sum_{q=0}^{p} \Phi_1^{2p-2q} S_f^q(r;u) + (p+1)\, \Phi_1^{2p} L_u \end{aligned}$$

because $L_u \geqq 0$ and $\Phi_1(r) \to +\infty$ for $r \to \infty$; q.e.d.

The _total inflection_ of f is defined by

$$\overset{+}{\mathfrak{S}}_f(r;u) = \sum_{p=0}^{n-1} \overset{+}{S}{}_f^p(r;u) \geqq 0$$

It is an increasing, continuous function of r.

Theorem 12.10. The Second Asymptotic Deficit Estimate. Assume (B1) - (B8). Then there exists a constant $C > 0$ such that for all $a \in \mathbb{P}(V^*)$ and for all $r > 0$ except a set of finite measure, the following estimate holds.

$$D_f(r,a) \leqq C\ \Phi(r)\ [\log^+ T_f(r) + \log^+ S_f^+(r;u) + \log^+ Z_u(r)$$

$$+ \log^+ Q_u(r) + \log^+ \Phi(r) + \log^+ \mathrm{Ric}_u^+(r)] + 2^n\ \overset{+}{S}_f(r,u)$$

Proof. For $r > 0$

$$\Phi_1(r) = \int_0^r \Phi(t)dt \leqq r\ \Phi(r)$$

Proposition 12.9 implies the existence of a constant $C_1 > 1$ such that

$$T_{f_p}(r) \leqq C_1(r\ \Phi(r))^{2p}(T_f(r) + \overset{+}{S}_f(r;u) + (\log^+ Z_u(r))^2 +$$

$$+ (Q_u(r))^2 + \mathrm{Ric}_u^+(r))$$

Hence

$$\log\left(1 + \frac{T_{f_p}(r)}{\Phi(r)}\right) \leqq \log 2 + \log^+ \frac{T_{f_p}(r)}{\Phi(r)} \leqq$$

$$\leqq \log 2 + \log C_1 + 2p \log^+ \Phi(r) + 2p \log r + \log 5$$

$$+ \log^+ T_f(r) + \log^+ \overset{+}{S}_f(r;u) + 2 \log^+ \log^+ Z_u(r)$$

$$+ 2 \log^+ Q_u(r) + \log^+ \mathrm{Ric}_u^+(r)$$

Take $s > 0$ with $A_f(s) > 0$. A constant $r_1 > r_0$ exists such that $T_f(r) \geqq s\, A_f(t)$ since T_f is unbounded. Hence

$$T_f(r) \geqq \int_s^r A_f(t)dt \geqq (r-s)\, A_f(s) \geqq r\, A_f(s) - T_f(r).$$

Hence $2\, T_f(r) \geqq r\, A_f(s) > 0$. A constant $C_2 > 1$ exists such that

$$\Phi(r) \log \left(1 + \frac{T_{f_p}(r)}{\Phi(r)}\right) \leqq$$

$$\leqq C_3\, \Phi(r)\, [\log^+ T_f(r) + \log^+ \overset{+}{S}_f(r,u) + \log^+ \Phi(r)$$

$$+ \log^+ \log^+ Z_u(r) + \log^+ Q_u(r) + \log^+ \mathrm{Ric}_u^+(r)]$$

Theorem 11.6 implies

$$D_f(r;a) \leqq 2^n\, \Phi(r) \sum_{p=0}^{n} \log \left(1 + \frac{T_{f_p}(r)}{\Phi(r)}\right) + 2^n \sum_{p=0}^{n-1} \overset{+}{S}{}_f^p(r)$$

$$+ n\, 2^{n-1}\, \Phi(r)\, [C_4 + \log^+ \Phi(r) + (m-1) \log^+ Z_u(r)]$$

where $C_4 > 0$ is a constant.

Here

$$\overset{+}{S}{}_f^p(r) = \frac{1}{2} \int_{G_r} \log^+ \frac{1}{h_p}\, \varphi^m = \frac{1}{2} \int_{G_r} \left(\log^+ \frac{\delta_u}{h_{up}}\right) \varphi^m$$

$$\leqq \frac{1}{2} \int_{G_r} \log^+ \frac{1}{h_{up}}\, \varphi^m + \frac{1}{2} \int_{G_r} \log^+ \delta_u\, \varphi^m$$

$$= \overset{+}{S}{}_f^p(r;u) + \frac{1}{2} \int_{G_r} \log^+ \delta_u \, \varphi^m$$

Here

$$\varphi^m = \delta_u \, \varphi_u^m \geqq \delta_u (u' \circ \tau)^m \, \varphi^m$$

Therefore $1 \geqq \delta_u(u' \circ \tau)^m \geqq 0$ almost everywhere on M. By continuity

$$0 \leqq \delta_u \leqq (u' \circ \tau)^{-m} \leqq u'(0)^{-m}$$

Hence a constant $C_5 > 0$ exists such that

$$\sum_{p=0}^{n-1} \overset{+}{S}{}_f^p(r) \leqq \overset{+}{S}_f(r;u) + C_5 \, \Phi(r)$$

Therefore a constant $C > 1$ exists such that

$$D_f(r;a) \underset{\bullet}{\leqq} C \, \Phi(r) \, [\log^+ T_f(r) + \log^+ \overset{+}{S}_f(r;u) + \log^+ \Phi(r)$$

$$+ \log^+ Z_u(r) + \log^+ Q_u(r) + \log^+ \mathrm{Ric}_u^+(r)]$$

where C_4 was majorized by $\log^+ T_f(r)$; q.e.d.

Uniform Estimates

The exceptional set may depend on a. Therefore it is worthwhile to have uniform estimates of $D_f(r;a)$ by data computed at θ where $\theta > 1$ is a constant.

Lemma 12.11. Assume (B1) - (B9). Let p be an integer with $0 \leqq p < n$. Take $r \geqq 0$. Then

$$\int_0^r \Omega_f^p(t;u)dt \leqq \frac{1}{2} A_{f_p}(r) + \frac{1}{2}(m-1)\, \Phi_1(r) \log Z_u(r) + Q_u(r).$$

Proof. If $r \leqq r_0$, then $G_r = \emptyset$ and the estimate holds trivially. Assume $r > r_0$, then

$$\int_0^r \Omega_f^p(t;u)dt = \frac{1}{4\pi} \int_0^r \Big(\int_{\Gamma_t} \log h_{up}\, d^c\tau \wedge \varphi^{m-1}\Big)dt$$

$$= \frac{1}{4\pi} \int_{G_r} \log h_{up}\, d\tau \wedge d^c\tau \wedge \varphi^{m-1}$$

$$= \frac{1}{2} \Phi_1(r) \int_{G_r} \log \frac{h_{up}}{\rho_u} \frac{d\tau \wedge d^c\tau \wedge \varphi^{m-1}}{2\pi\ \Phi_1(r)}$$

$$+ \frac{1}{4\pi} \int_{G_r} \log \rho_u\, d\tau \wedge d^c\tau \wedge \varphi^{m-1}$$

$$\leqq \frac{1}{2} \Phi_1(r) \log \Big[\int_{G_r} \frac{h_{up}}{\rho_u} \frac{d\tau \wedge d^c\tau \wedge \varphi^{m-1}}{2\pi\ \Phi_1(r)}\Big] + Q_u(r)$$

$$\leqq \frac{1}{2} \Phi_1(r) \log \Big(\frac{1}{\Phi_1(r)} \int_{G_r} H_p\Big) + Q_u(r)$$

$$\leqq \frac{1}{2} \Phi_1(r) \log \Big[\frac{(Z_u(r))^{m-1}}{\Phi_1(r)} A_{f_p}(r)\Big] + Q_u(r)$$

$$\leqq \frac{1}{2} A_f(r) + \frac{m-1}{2} \Phi_1(r) \log Z_u(r) + Q_u(r) \qquad \text{q.e.d.}$$

Theorem 12.12. Assume (B1) - (B9). Let p be an integer with $0 \leqq p < n$. Take $\theta > 1$ and $r > 1$. Then

$$T_{f_p}(r) \leqq \left(\frac{2n\ \theta}{\theta-1}\right)^{2n} [T_f(\theta r) + (m-1)\ \Phi(\theta r) \log^+ Z_u(\theta r)$$

$$+ Q_u(\theta r) + \mathrm{Ric}_u^+(\theta r) + \overset{+}{S}_f(\theta r;u)].$$

Proof. Plücker's difference formula implies

$$T_{f_{p+1}}(r) \leqq 2\ T_{f_p}(r) + \Omega_f^p(r;u) + \overset{+}{S}{}_f^p(r;u) + \mathrm{Ric}_u^+(r)$$

Double integration yields

$$\int_0^r \int_0^x T_{f_{p+1}}(r)\ dt\ dx \leqq$$

$$\leqq r^2\ T_{f_p}(r) + \int_0^r \int_0^x \Omega_f^p(t;u)\ dtdx + \frac{r^2}{2}\ \overset{+}{S}{}_f^p(r,u) + \frac{r^2}{2}\ \mathrm{Ric}_u^+(r)$$

Lemma 12.11 implies

$$\int_0^r \int_0^x \Omega_f^p(t;u)\ dt\ dx \leqq$$

$$\leqq \frac{1}{2}\ T_{f_p}(r) + \frac{1}{2}\ (m-1) r\ \Phi_1(r) \log^+ Z_u(r) + r\ Q_u(r)$$

If $\theta > 1$, then

$$\int_0^{\theta r} \int_0^x T_{f_{p+1}}(t)dt\ dx \geqq \int_r^{\theta r} \int_r^x T_{f_{p+1}}(t)dt\ dx \geqq$$

$$\geqq\ T_{f_{p+1}}(r) \int_r^{\theta r} \int_r^x dt\ dx = T_{f_{p+1}}(r)\ \frac{r^2}{2}\ (\theta-1)^2$$

$$\Phi_1(\theta r) = \int_0^{\theta r} \Phi(t)\ dt \leqq \theta r\ \Phi\ (\theta r)$$

If $\theta > 1$ and $r > 1$, these estimates imply

$$T_{f_{p+1}}(r) \leqq \frac{1}{(\theta-1)^2} [2\theta^2 T_{f_p}(\theta r) + \frac{1}{r^2} T_{f_p}(\theta r) +$$

$$+ (m-1)\ \theta^2\ \Phi(\theta r)\ \log^+ Z_u(\theta r) + \frac{2\theta}{r} Q_u(\theta r)$$

$$+ \theta^2 \overset{+}{S}{}^p_f(\theta r;u) + \theta^2\ \mathrm{Ric}^+_u(\theta r)]$$

$$\leqq \frac{3\theta^2}{(\theta-1)^2} [T_{f_p}(\theta r) + (m-1)\ \Phi(\theta r)\ \log^+ Z_u(\theta r) + Q_u(\theta r)$$

$$+ \mathrm{Ric}^+_u(\theta r) + \overset{+}{S}{}^p_f(\theta r;u)]$$

which implies

$$T_{f_1}(r) \leqq (\frac{2\theta}{\theta-1})^2 [T_f(\theta r) + (m-1)\ \Phi(\theta r)\ \log^+ Z_u(\theta r) +$$

$$+ Q_u(\theta r) + \mathrm{Ric}^+_u(\theta r) + \overset{+}{S}{}^0_f(\theta r;u)]$$

Now, the following estimate is claimed

$$T_{f_p}(r) \leqq (\frac{2\theta}{\theta-1})^{2p} [T_f(\theta^p r) + (m-1)\ \Phi(\theta^p r)\ \log^+ Z_u(\theta^p r) +$$

$$+ Q_u(\theta r) + \mathrm{Ric}^+_u(\theta r) + \sum_{q=0}^{p-1} \overset{+}{S}{}^q_f(\theta^p r;u)]$$

The statement is correct for $p = 1$. Assume, that it is proven for p then it shall be proved for $p + 1$ if $p+1 < n$.

$$T_{f_{p+1}}(r) \leqq \frac{3\theta^2}{(\theta-1)^2} [(\frac{2\theta}{\theta-1})^{2p} (T_f(\theta^{p+1} r) + Q_u(\theta^{p+1} r)$$

$$+ (m-1)\ \Phi(\theta^{p+1}r)\ \log^+ Z_u(\theta^{p+1}r) + \sum_{q=0}^{p-1} \overset{+}{S}{}_f^q(\theta^{p+1}r;u)$$

$$+ \mathrm{Ric}_u^+(\theta^{p+1}r) + Q_u(\theta r) + \overset{+}{S}{}_f^p(\theta r;u)$$

$$+ \mathrm{Ric}_u(\theta r)]$$

Here $\theta < \theta^{p+1}$ and

$$2 \leqq \frac{3\theta^2}{(\theta-1)^2}\left[\left(\frac{2\theta}{\theta-1}\right)^{2p} + 1\right] \leqq (3\cdot 2^{2p} + 1)\left(\frac{\theta}{\theta-1}\right)^{2p+2} \leqq \left(\frac{2\theta}{\theta-1}\right)^{2p+2}$$

Therefore

$$T_{f_{p+1}}(r) \leqq \left(\frac{2\theta}{\theta-1}\right)^{2p+2} [T_f(\theta^{p+1}r) + (m-1)\ \Phi(\theta^{p+1}r)\log^+ Z_u(\theta^{p+1}r)$$

$$+ Q_u(\theta^{p+1}r) + \sum_{q=0}^{p} \overset{+}{S}{}^q(\theta^{p+1}r;u) + \mathrm{Ric}_u^+(\theta^{p+1}r)]$$

Hence the claim is proved by induction. Observe $\theta^p \leqq \theta^n$ and

$$\frac{\theta}{\theta-1} = \frac{\theta^n + \theta^{n-1} + \ldots + \theta}{\theta^n - 1} \leqq \frac{n\theta^n}{\theta^n - 1}$$

Now, take any $\theta > 1$. Define $\theta_1 = \theta^{\frac{1}{n}}$. Then $\theta = \theta_1^n$. Use the claim for $p = n$ and θ_1. Then $\left(\frac{2\theta_1}{\theta_1-1}\right)^{2p} \leqq \left(\frac{2n\theta}{\theta-1}\right)^{2n}$. Hence

$$T_{f_p}(r) \leqq \left(\frac{2n\theta}{\theta-1}\right)^{2n} [T_f(\theta r) + (m-1)\ \Phi(\theta r) \cdot \log^+ Z_u(\theta r)$$

$$+ Q_u(\theta r) + \sum_{q=0}^{p-1} \overset{+}{S}{}_f^q(\theta r;u) + \mathrm{Ric}_u^+(\theta r)]$$

q.e.d.

Theorem 12.13. Second Uniform Deficit Estimate. Assume (B1) - (B9). Then there exists a constant $C > 0$ depending on m, n and $u'(0)$ only, such that for each pair of numbers θ and r with $1 < \theta \leqq 2 < r$ and for each $a \in \mathbb{P}(V^*)$ the following estimate holds

$$D_f(r;a) = 2^{2n+1}\,\Phi(r)\,[\log^+ T_f(\theta r) + \log^+ \Phi(\theta r)$$

$$+ (m-1)\log^+ Z_u(\theta r) + \log^+ Q_u(\theta r)$$

$$+ \log^+ \overset{+}{S}_f(\theta r;u) + \log^+ \frac{\theta}{\theta-1} + C]$$

$$+ 2^n\,(1 + \overset{+}{S}_f(r;u)).$$

Proof. By Theorem 11.5, there exists a constant $C_1 > 0$ such that

$$D_f(r,a) \leqq 2^n\,\Phi(r) \sum_{p=0}^{n} \log^+ \frac{T_{f_p}(\theta r) + \Phi(\theta r)}{r(\theta-1)\,\Phi(r)} + 2^n \sum_{p=0}^{n-1} \overset{+}{S}{}^p_f(r)$$

$$+ \Phi(r)\,[C_1 + n(m-1)2^{n-1}\log^+ Z_u(\theta r)]$$

if $r \geqq r_0$ and $\theta > 1$. Take $r > \mathrm{Max}(r_0, 2)$ and $\theta > 1$. Then

$$\Phi(r)\log^+ \frac{T_{f_p}(\theta r) + \Phi(\theta r)}{r\,\Phi(r)\,(\theta-1)} \leqq$$

$$\leqq \Phi(r)\,[\log^+ \frac{1}{\Phi(r)} + \log\frac{1}{(\theta-1)} + \log 2 + \log^+ T_{f_p}(\theta r) + \log^+\Phi(\theta r)]$$

$$\leqq 1 + \Phi(r)\,[(2n+1)\log\frac{\theta}{\theta-1} + 2n\log 2n + \log 10 +$$

$$+ \log^+ T_f(\theta^2 r) + \log^+((m-1)\log^+ Z_u(\theta^2 r)) + 2\log^+ \Phi(\theta^2 r)$$

$$+ \log^+ Q_u(\theta^2 r) + \log^+ \mathrm{Ric}_u^+(\theta^2 r) + \log^+ \overset{+}{S}_f(\theta^2 r;u)]$$

As shown in the proof of Theorem 12.10, a constant $C_2 > 0$ exists such that

$$\sum_{p=0}^{n-1} \overset{+}{S}{}_f^p(r) \leqq \overset{+}{S}_f(r;u) + C_2\,\Phi(r)$$

The constant C_2 depends on m and u'(0) only. Therefore there exists a constant C_3 which depends on n, m and u'(0) only such that

$$D_f(r;a) \leqq 2^n(1 + \overset{+}{S}_f(r;u)) +$$

$$2^n\,\Phi(r)\,[\log^+ T_f(\theta^2 r) + (2n+1)\log\frac{\theta}{\theta-1} + 2\log^+\Phi(\theta^2 r)$$

$$+ n(m-1)\log^+ Z_u(\theta^2 r) + \log^+ Q_u(\theta^2 r) + \log^+ \mathrm{Ric}_u^+(\theta^2 r)$$

$$+ \log^+ \overset{+}{S}_f(\theta^2 r;u) + C_3]$$

Here $n \leqq 2n + 1 < 2^{n+1}$ and

$$\frac{\theta}{\theta-1} = \frac{\theta^2+\theta}{\theta^2-1} \leqq \frac{2\theta^2}{\theta^2-1}$$

Take any θ with $1 < \theta \leqq 2$. Define $\theta_1 = \theta^{\frac{1}{2}}$. Then $1 < \theta_1 < \theta_1^2 = \theta \leqq 2$. Apply the last estimate to θ_1. Then a constant $C > 0$ exists which depends on n, m and u'(0) only such that

$$D_f(r;a) \leqq 2^{2n+1}\,\Phi(r)\,[\log^+ T_f(\theta r) + \log^+ \Phi(\theta r) + \log^+ Q_u(\theta r)$$

$$+ (m-1)\log^+ Z_u(\theta r) + \log^+ \mathrm{Ric}_u^+(\theta r) + \log^+ \overset{+}{S}_f(\theta r;u) + C]$$

$$+ 2^n\,[1 + \overset{+}{S}_f(r;u)]$$

For all $r > \mathrm{Max}\,(r_0, 2)$. If $r_0 \geqq 2$ and $r_0 \geqq r \geqq 2$, the inequality is trivial true because $D_f(r;a) = 0$. Hence the estimate holds for all $r \geqq 2$; q.e.d.

XIII. STEADY MAPS

The counter example of Cornalba and Shiffman [9] has shown that T_f alone cannot majorize the deficit $D_f(r;a)$. The main value of the deficit estimate is to expose explicitly those invariants of M and f responsible for this fact. In the classical theory, $\Phi(r)$ is the volume of a ball, hence $\log \Phi(r)$ grows as log r. For transcendental maps f the characteristic grows much faster. Hence it would be natural to assume that f grows so fast as to overcome $\Phi(r)$, $Q_u(r)$ and $\mathrm{Ric}_u^+(r)$ which depend only on the geometry of M. Such an assumption will be made at the end of this section; however it is not permissible for an application to Bezout's problem where the growth of $\Phi(r)$ and $T_f(r)$ will be about the same.

The examples of Griffiths [16] and Cornalba-Shiffman [9] show that $\overset{+}{S}_f(r;u)$ may grow much quicker than $T_f(r)$. Also it should be noted that $\overset{+}{S}_f(r;u)$ depends on the selection of B. Hence, only some radical and perhaps artificial assumption will majorize $\overset{+}{S}_f$. Here a suggestion of Griffiths [16] will be investigated and some geometric interpretation of the assumption will be given. The assumption itself is radical and may be violated in many cases.

Assume (B1) - (B9). The map f is said to be <u>steady for B</u> if and only if there exists a constant $C > 0$ such that $h_{up}(z) \geqq C > 0$ for all $z \in M - I_{f_p}$ and for all $p = 0,1,\dots,n-1$.

The following <u>additional general assumption</u> will be made.

(B10) Assume that f is steady for B.

<u>Proposition 13.1</u>. Assume (B1) - (B10). Then the pth stationary divisor v_p is the zero divisor for $p = 0,1,\dots,n-1$, that is $v_p \equiv 0$. Moreover $I_{f_p} \subseteq I_f$ for $p = 1,\dots,n-1$. Especially, if f is holomorphic, then all associated maps f_p are holomorphic.

Proof. Take $x_0 \in M$. Take a patch $\alpha \in \mathfrak{T}_M$ with $x_0 \in U_\alpha$ such that $\alpha: U_\alpha \to U'_\alpha$ and $U'_\alpha = \mathbb{C}^m(1)$ is the unit ball. Then an irreducible representation $\mathfrak{w}: U_\alpha \to V$ of f on U_α exists. Define $\mathfrak{w}_p = \mathfrak{w}_{p\alpha}$ for $p = 0,1,\dots,n$. Then a holomorphic function $D_p \not\equiv 0$ on U_α exists such that $\mathfrak{w}_p = D_p \mathfrak{y}_p$ where $\mathfrak{y}_p$ is an irreducible representation of f_p on U_α. According to Lemma 8.1 v_p is the divisor of $E_p = D_{p+1} D_{p-1} D_p^{-2}$ on U_α. Here $D_{-1} = 1$. Hence E_p is a holomorphic function on U_α. Define

$$\upsilon_\alpha = (\tfrac{i}{2\pi})^m\, d\alpha_1 \wedge d\bar{\alpha}_1 \wedge \cdots \wedge d\alpha_m \wedge d\bar{\alpha}_m$$

Then $H_p = H_p^\alpha\, \upsilon_\alpha$ and $\varphi_u^m = \varphi_\alpha\, \upsilon_p$ on U_α where $\varphi_\alpha > 0$ is a function of class C^∞ on U_α. Also $H_p^\alpha = h_{up}\, \varphi_\alpha$ on U_α. A constant $C > 0$ exists such that $H_p^\alpha \geqq C\, \varphi_\alpha$ on U_α. Hence

$$|E_p|^2\, |\mathfrak{y}_{p+1}|^2\, |\mathfrak{y}_{p-1}|^2 \geqq C\, \varphi_\alpha\, |\mathfrak{y}_p|^2$$

on U_α with $\mathfrak{y}_{-1} = 1$. Because $\mathfrak{y}_q^{-1}(0)$ has atmost dimension m-2, the function E_p is not zero except perhaps on an anlytic set of codimension 2. Therefore $E_p(x) \neq 0$ for all $x \in U_\alpha$. Hence $v_p(x) = 0$ for all $x \in U_\alpha$ and $p = 0,1,\ldots,n-1$. Now, take $x \in U_\alpha \cap I_{f_{p+1}}$ then $\mathfrak{y}_{p+1}(x) = 0$. Hence $\mathfrak{y}_p(x) = 0$. Therefore $x \in U_\alpha \cap I_{f_p}$. Consequently $I_{f_n} \subseteq I_{f_{n-1}} \subseteq \ldots \subseteq I_f$;

q.e.d.

Lemma 13.2. Assume (B1) - (B10). Then there exists a constant $C_0 > 0$ such that

$$\overset{+}{S}_f(r;u) \leqq C_0\, \Phi(r) \qquad \text{if } r \geqq 0$$

Proof. A constant C with $0 < C < 1$ exists such that $h_{up} \geqq C > 0$. Then

$$\overset{+}{S}_f(r;u) = \sum_{p=0}^{n-1} \frac{1}{2} \int_{G_r} \log^+ \frac{1}{h_{up}}\, \varphi^m \leqq (\tfrac{n}{2} \log \tfrac{1}{C}) \int_{G_r} \varphi^m = C_0 \Phi(r)$$

where $\frac{n}{2} \log \frac{1}{C} = C_0 > 0$ is constant,

q.e.d.

Lemma 13.3. Assume (B1) - (B10). Then a constant $c > 0$ exists such that

$$C\ \Phi(r)(u'(r))^m \leqq (Z_u(r))^{m-1}\ A_{f_p}(r)$$

for all $r \geqq 0$ and $p = 0,1,\ldots,n-1$.

Proof. A constant $C > 0$ exists such that $h_{up} \geqq C > 0$ on $M - I_{f_p}$. Hence $H_p = h_{up}\ \varphi_u^m \geqq C\ \varphi_u^m$ on $M - I_{f_p}$. Lemma 12.4 implies

$$(Z_u(r))^{m-1} A_{f_p}(r) \geqq \int_{G_r} H_p \geqq C \int_{G_r} \varphi_u^m = C(u'(r))^m \Phi(r)$$

q.e.d.

Now, the deficit estimates can be formulated for steady maps.

Theorem 13.4. Third Asymptotic Deficit Estimate. Assume (B1) - (B10). Then there exists a constant $C > 0$ such that for all $a \in \mathbb{P}(V^*)$ and all $r > 0$ except for a set of finite measure the following estimate holds

$$D_f(r;a) \underset{\bullet}{\leqq} C\ \Phi(r)\ [\log^+ T_f(r) + \log^+ Z_u(r) + \log^+ Q_u(r)$$

$$+ \log^+ \Phi(r) + \log^+ \mathrm{Ric}_u^+(r)]$$

Proof. Observe T_f is not bounded. Hence constants C_0, C_1 and C_2 exist such that

$$\log^+ \overset{+}{S}_f(r;u) \underset{\bullet}{\leqq} \log^+\Phi(r) + \log^+ C_0 \leqq \log^+ \Phi(r) + \log^+ T_f(r)$$

$$2^n\ \overset{+}{S}_f(r,u) \leqq C_1\ \Phi(r) \leqq \Phi(r)\ \log^+ T_f(r)$$

Hence Theorem 12.10 implies Theorem 13.4 immediately.

The same estimates and 12.13 imply

Theorem 13.5. Third Uniform Deficit Estimate.

Assume (B1) - (B10). Then constants $C_1 > 0$ and $C_2 > 0$ exist such that for all $a \in \mathbb{P}(V^*)$ and all real numbers r and θ with $1 < \theta \leqq 2 < r$ the following estimate holds

$$D_f(r;a) \leqq 2^{2n+1}\,\Phi(r)\,[\log^+ T_f(\theta r) + C_1 \log^+ \Phi(\theta r)$$

$$+ (m-1)\log^+ Z_u(\theta r) + \log^+ Q_u(\theta r)$$

$$+ \log^+ \mathrm{Ric}_u^+(\theta r) + \log\frac{\theta}{\theta-1} + C_2] + 2^n$$

Now, this program is interrupted to bring the following geometric interpretations of steadiness for B. Assume (B1) - (B10). At first some notations shall be assembled.

General Notations

Let $\hat{\mathfrak{T}}_M$ be the set of all pairs $(\alpha, \mathfrak{v})$ where $\alpha \in \mathfrak{T}_M$ is a patch on M and where $\mathfrak{v}: U_\alpha \to V'$ is an irreducible representation of f on U_α. This set is indexed by an index set Λ and written as a family $\hat{\mathfrak{T}}_M = \{(\alpha_\lambda, \mathfrak{v}_\lambda)\}_{\lambda \in \Lambda}$. The map $\lambda \to (\alpha_\lambda, \mathfrak{v}_\lambda)$ can be taken one-to-one. For $\lambda \in \Lambda$ write

$$U_\lambda = U_{\alpha_\lambda} \qquad U'_\lambda = U'_{\alpha_\lambda}$$

$$\alpha_\lambda = (\alpha_{\lambda 1}, \ldots, \alpha_{\lambda m})$$

$$\zeta_\lambda = d\alpha_{\lambda 1} \wedge \cdots \wedge d\alpha_{\lambda m}$$

$$\zeta_\lambda^{\otimes q} = \zeta_\lambda \otimes \ldots \otimes \zeta_\lambda \quad \text{(q-times)}$$

$$\upsilon_\lambda = \left(\frac{i}{2\pi}\right)^m d\alpha_{\lambda 1} \wedge d\bar{\alpha}_{\lambda 1} \wedge \cdots \wedge d\alpha_{\lambda m} \wedge d\bar{\alpha}_{\lambda m}$$

$$\upsilon_\lambda = i_0 \zeta_\lambda \wedge \bar{\zeta}_\lambda$$

$$\zeta_\lambda^\nu = (-1)^{\nu-1} d\alpha_{\lambda 1} \wedge \cdots \wedge d\alpha_{\lambda \nu-1} \wedge d\alpha_{\lambda \nu+1} \wedge \cdots \wedge d\alpha_{\lambda m}$$

$$B = \sum_{\nu=1}^{m} b_{\lambda\nu} \zeta_\lambda^\nu \quad \text{on } U_\lambda$$

$$\mathfrak{w}_\lambda^{(\mu)} = \mathfrak{w}_{\alpha_\lambda}^{(\mu)} \quad \text{on } U_\lambda$$

$$\mathfrak{w}_{\lambda q} = \mathfrak{w}_\lambda \wedge \mathfrak{w}'_\lambda \wedge \cdots \wedge \mathfrak{w}_\lambda^{(q)} \quad \text{and} \quad \mathfrak{w}_{\lambda,-1} = 1$$

$$H_p^\lambda = \frac{|\mathfrak{w}_{\lambda p-1}|^2 \, |\mathfrak{w}_{\lambda p+1}|^2}{|\mathfrak{w}_{\lambda p}|^4} \quad \text{on } U_\lambda - \mathfrak{w}_{\lambda p}^{-1}(0)$$

$$H_p = H_p^\lambda \upsilon_\lambda \quad \text{on } U_\lambda - (\mathfrak{w}_{\lambda p})^{-1}(0)$$

$$\varphi_u^m = \varphi_\lambda \upsilon_\lambda \quad \text{on } U_\lambda$$

$$H_p^\lambda = \varphi_\lambda h_{up} \quad \text{on } U_\lambda - (\mathfrak{w}_{\lambda p}^{-1})(0)$$

Here ζ_λ is a holomorphic frame of the canonical bundle K of M over U_λ and $\zeta_\lambda^{\otimes q}$ is a holomorphic frame of the tensor product K^q. Define

$$U_{\lambda_0 \lambda_1 \ldots \lambda_p} = U_{\lambda_0} \cap \cdots \cap U_{\lambda_p}$$

$$\Lambda[p] = \{(\lambda_0, \ldots, \lambda_p) \mid U_{\lambda_0 \lambda_1, \ldots, \lambda_p} \neq \emptyset\}$$

If $(\lambda, \mu) \in \Lambda[1]$, then

$$\zeta_\lambda = \Delta_{\lambda\mu}\, \zeta_\mu \qquad\qquad \mathfrak{w}_\lambda = g_{\lambda\mu}\, \mathfrak{w}_\mu$$

$$\mathfrak{w}_{\lambda q} = (g_{\lambda\mu})^{q+1} (\Delta_{\mu\lambda})^{\frac{q(q+1)}{2}} \mathfrak{w}_{\mu q}$$

on $U_{\lambda\mu}$. Here $\Delta_{\lambda\mu}$ and $g_{\lambda\mu}$ are holomorphic functions on $U_{\lambda\mu}$ with $\Delta_{\lambda\mu}(x) \neq 0 \neq g_{\lambda\mu}(x)$ for all $x \in U_{\lambda\mu}$. Hence

$$\mathfrak{w}_{\lambda q}^{-1}(0) \cap U_\mu = \mathfrak{w}_{\mu q}^{-1}(0) \cap U_\lambda$$

Therefore a thin analytic subset I_q of M exists such that $U_\lambda \cap I_q = \mathfrak{w}_{\lambda q}^{-1}(0) \cap U_\mu$. Obviously $I_q \supseteq I_{f_q}$. Define

$$I = I_0 \cup \ldots \cup I_n \qquad\qquad M^* = M - I$$

Observe that $\{\Delta_{\lambda\mu}\}_{(\lambda,\mu) \in \Lambda[1]}$ is a basic cocycle of the canonical bundle and that $\{g_{\lambda\mu}\}_{(\lambda,\mu) \in \Lambda[1]}$ is a basic cocycle of the (extended) pullback $L = f^*(S(V))$ of the hypersection bundle.

1. Interpretation

Steadiness means that f is bounded away from inflection and higher contact with its tangent plane. All stationary divisors are zero and even $H_p \geqq C\, \varphi_u^m$, meaning there is not even an inflection at "infinity". This interpretation is due to Griffiths [16].

2. Interpretation

Steadiness for B will be shown to mean that the invariant flag ratio is uniformly bounded below, that is, the flag does not collapse.

Let $GL(V)$ be the set of all $\mathfrak{w} = (\mathfrak{w}_0, \ldots, \mathfrak{w}_n)$ with $\det \mathfrak{w} \neq 0$. Hence $GL(V)$ is the set of all bases of V. It is an open, connected, dense subset of $V^{n+1} = V \times \ldots \times V$. To $\mathfrak{w} \in GL(V)$ a flag

$$\{\mathfrak{w}_0, \mathfrak{w}_0 \wedge \mathfrak{w}_1, \ldots, \mathfrak{w}_0 \wedge \ldots \wedge \mathfrak{w}_n\}$$

is assigned called the flag of $\mathfrak{w}$. A base $\mathfrak{v} = (\mathfrak{v}_0, \ldots, \mathfrak{v}_n) \in GL(V)$ is called a flag frame to $\mathfrak{w}$ if and only if

$$(\mathfrak{v}_\mu \mid \mathfrak{v}_\nu) = 0 \quad \text{if} \quad 0 \leqq \mu < \nu \leqq n$$

$$\mathfrak{v}_0 \wedge \ldots \wedge \mathfrak{v}_p = \mathfrak{w}_0 \wedge \ldots \wedge \mathfrak{w}_p \quad \text{for } p = 0, 1, \ldots, n.$$

Lemma 13.6. Take $\mathfrak{w} = (\mathfrak{w}_0, \ldots, \mathfrak{w}_n) \in GL(V)$. Then one and only one flag frame $\mathfrak{v} = \mathfrak{v}(\mathfrak{w})$ to $\mathfrak{w}$ exists. The map defined by $\mathfrak{w} \to \mathfrak{v}(\mathfrak{w})$ is of class C^∞ on $GL(V)$. Moreover

$$|\mathfrak{v}_p| = \frac{|\mathfrak{w}_0 \wedge \ldots \wedge \mathfrak{w}_p|}{|\mathfrak{w}_0 \wedge \ldots \wedge \mathfrak{w}_{p-1}|} \qquad \text{if } 1 \leqq p \leqq n$$

$$|\mathfrak{v}_0| = |\mathfrak{w}_0|$$

Proof. a) Existence. Take $\mathfrak{v}_0 = \mathfrak{a}_0$. Assume $\mathfrak{v}_0, \ldots, \mathfrak{v}_p$ are already constructed as functions of class C^∞ such that $(\mathfrak{v}_\mu | \mathfrak{v}_\nu) = 0$ for $0 \leqq \mu \leqq p$ and $0 \leqq \nu \leqq p$ if $\mu \neq \nu$ and such that $\mathfrak{v}_0 \wedge \cdots \wedge \mathfrak{v}_\mu = \mathfrak{a}_0 \wedge \cdots \wedge \mathfrak{a}_\mu$ for $0 \leqq \mu \leqq p < n$. Define the vector function $\mathfrak{v}_{p+1}$ of class C^∞ on GL(V) by

$$\mathfrak{v}_{p+1} = \mathfrak{a}_{p+1} - \sum_{\mu=0}^{p} (\mathfrak{a}_{p+1} | \mathfrak{v}_\mu) | \mathfrak{v}_\mu |^{-2} \mathfrak{v}_\mu$$

Then $(\mathfrak{v}_{p+1} | \mathfrak{v}_\mu) = 0$ if $0 \leqq \mu \leqq p$. Also

$$\mathfrak{v}_0 \wedge \cdots \wedge \mathfrak{v}_{p+1} = \mathfrak{v}_0 \wedge \cdots \wedge \mathfrak{v}_p \wedge \mathfrak{a}_{p+1} = \mathfrak{a}_0 \wedge \cdots \wedge \mathfrak{a}_{p+1}.$$

The existence is proved by induction

b) Uniqueness. Assume $(\mathfrak{w}_0, \ldots, \mathfrak{w}_n)$ is a solution for an $\mathfrak{a} \in$ GL(V). Write $\mathfrak{v}_\mu = \mathfrak{v}_\mu(\mathfrak{a})$. Then $\mathfrak{w}_0 = \mathfrak{a}_0 = \mathfrak{v}_0$. Assume $\mathfrak{w}_0 = \mathfrak{v}_0, \ldots, \mathfrak{w}_p = \mathfrak{v}_p$ for some $p < n$. Then

$$\mathfrak{w}_{p+1} = a_0 \mathfrak{v}_0 + \cdots + a_{p+1} \mathfrak{v}_{p+1}$$

with $a_\mu = (\mathfrak{w}_{p+1} | \mathfrak{v}_\mu)$ for $\mu = 0, 1, \ldots, p+1$. If $0 \leqq \mu \leqq p$, then $a_\mu = (\mathfrak{w}_{p+1} | \mathfrak{v}_\mu) = (\mathfrak{w}_{p+1} | \mathfrak{w}_\mu) = 0$. Hence $a = a_{p+1}$ and $\mathfrak{w}_{p+1} = a\, \mathfrak{v}_{p+1} \neq 0$. Therefore

$$0 \neq \mathfrak{a}_0 \wedge \cdots \wedge \mathfrak{a}_{p+1} = \mathfrak{w}_0 \wedge \cdots \wedge \mathfrak{w}_{p+1} = \mathfrak{v}_0 \wedge \cdots \wedge \mathfrak{v}_p \wedge \mathfrak{w}_{p+1}$$

$$= a\, \mathfrak{v}_0 \wedge \cdots \wedge \mathfrak{v}_{p+1} = a\, \mathfrak{a}_0 \wedge \cdots \wedge \mathfrak{a}_{p+1}$$

Therefore a = 1. The uniqueness is proved by induction.

Observe $|\mathfrak{v}_0 \wedge \cdots \wedge \mathfrak{v}_p| = |\mathfrak{v}_0| \cdots |\mathfrak{v}_p|$ and $|\mathfrak{w}_0 \wedge \cdots \wedge \mathfrak{w}_p| = |\mathfrak{v}_0 \wedge \cdots \wedge \mathfrak{v}_p|$. If $1 \leqq p \leqq n$, then

$$\frac{|\mathfrak{w}_0 \wedge \cdots \wedge \mathfrak{w}_p|}{|\mathfrak{w}_0 \wedge \cdots \wedge \mathfrak{w}_{p-1}|} = \frac{|\mathfrak{v}_0| \cdots |\mathfrak{v}_p|}{|\mathfrak{v}_0| \cdots |\mathfrak{v}_{p-1}|} = |\mathfrak{v}_p|$$

Obviously $|\mathfrak{v}_0| = |\mathfrak{w}_0|$; q.e.d.

The interpretation will take place only on M*. If $\lambda \in \Lambda$ and $z \in U_\lambda \cap M^*$, then $\mathfrak{w}_\lambda(z), \mathfrak{w}_\lambda'(z), \ldots, \mathfrak{w}_\lambda^{(n)}(z)$ is a base of V. Let $\mathfrak{e}_{\lambda 0}(z), \ldots, \mathfrak{e}_{\lambda n}(z)$ be the flag frame associated to this base. Hence $\mathfrak{e}_{\lambda p}\colon U_\lambda \cap M^* \to V - \{0\}$ is a vector function of class C^∞ such that

$$(\mathfrak{e}_{\lambda p} | \mathfrak{e}_{\lambda q}) = 0 \quad \text{if } p \neq q \qquad \text{on } U_\lambda - M^*$$

$$\mathfrak{e}_{\lambda 0} \wedge \cdots \wedge \mathfrak{e}_{\lambda p} = \mathfrak{w}_{\lambda p} \qquad \text{on } U_\lambda - M^*$$

<u>Lemma 13.7</u>. Take $(\lambda, \mu) \in \Lambda[1]$. Then

$$\mathfrak{e}_{\lambda p} = g_{\lambda\mu}\, \Delta_{\mu\lambda}^{p}\, \mathfrak{e}_{\mu p} \qquad \text{on } U_{\lambda\mu}.$$

Proof. On $U_{\lambda\mu}$ define $\mathfrak{y}_{\lambda p} = g_{\lambda\mu}\, \Delta_{\mu\lambda}^{p}\, \mathfrak{e}_{\mu p}$. Then $(\mathfrak{y}_{\lambda p} | \mathfrak{y}_{\lambda q}) = 0$ if $0 \leqq p < q \leqq n$. Also

$$\mathfrak{y}_{\lambda 0} \wedge \cdots \wedge \mathfrak{y}_{\lambda p} = g_{\lambda\mu}^{p+1} (\Delta_{\mu\lambda})^{\frac{p(p+1)}{2}}\, \mathfrak{e}_{\mu 0} \wedge \cdots \wedge \mathfrak{e}_{\mu p}$$

$$= g_{\lambda\mu}^{p+1} (\Delta_{\mu\lambda})^{\frac{p(p+1)}{2}}\, \mathfrak{w}_{\mu p} = \mathfrak{w}_{\lambda p}$$

Hence uniqueness implies $\mathfrak{y}_{\lambda p} = \mathfrak{E}_{\lambda p}$ for $p = 0,1,\ldots,n$; q.e.d.

Recall that $\mathfrak{w}_\lambda$ can be considered as a holomorphic frame of L over U_λ. Let $\mathfrak{w}_\lambda^*$ be the dual frame of the dual bundle L^*. If $w \in \mathbb{C}$, then $\mathfrak{w}_\lambda^*(x)\ (w\,\mathfrak{w}_\lambda(x)) = w$. Also $\mathfrak{w}_\lambda^* = g_{\mu\lambda}\, \mathfrak{w}_\mu^*$ on $U_{\lambda\mu}$ if $(\lambda,\mu) \in \Lambda[1]$. (See section 4). Then

$$\mathfrak{E}_{\lambda p} \otimes \mathfrak{w}_\lambda^* \otimes \zeta_\lambda^{\otimes p} = g_{\lambda\mu}\, \Delta_{\mu\lambda}^p \quad g_{\mu\lambda} \quad \Delta_{\lambda\mu}^p \quad \mathfrak{E}_{\mu p} \otimes \mathfrak{w}_\mu^* \otimes \zeta_\mu^{\otimes p}$$

$$= \mathfrak{E}_{\mu p} \otimes \mathfrak{w}_\mu^* \otimes \zeta_\mu^{\otimes p}$$

on $U_{\lambda\mu} \cap M^*$ if $(\lambda,\mu) \in \Lambda[1]$. Let $V_M = M \times V$ be the trivial bundle. One and only one section

$$\mathfrak{E}_p\colon M^* \to V_M \otimes L^* \otimes K^p$$

of class C^∞ exists such that $\mathfrak{E}_p | U_\lambda \cap M^* = \mathfrak{E}_{\lambda p}$ for all $\lambda \in \Lambda$. Hence $\mathfrak{E}_0, \mathfrak{E}_1, \ldots, \mathfrak{E}_p$ can be considered to be the flag frame of an "invariant" flag.

The hermitian metric on V defines an hermitian metric along the fibers of V_M. Because $L|M^*$ is a subbundle of $V_M|M^* = V_{M^*}$, this restricts to a hermitian metric along the fibers of $L|M^*$. It induces a dual metric on $L^*|M^*$ such that $|\mathfrak{w}_\lambda^*|\ |\mathfrak{w}_\lambda| = 1$ on $U_\lambda \cap M^*$. On K, a hermitian metric along the fibers of κ was defined by $\kappa(\zeta_\lambda) = \frac{1}{\varphi_\lambda}$. A hermitian metric κ^p along the fibers is defined by

$$\kappa^p(\zeta_\lambda^{\otimes p}) = \frac{1}{\varphi_\lambda^p}$$

The product of these three metrics defines a hermitian metric along the fibers of $V_M \otimes L^* \otimes K$ over M^* whose length is denoted by $\|\ \|$. Then

$$\| \mathfrak{z}_p \|^2 = \frac{|\mathfrak{z}_{\lambda p}|^2}{|\mathfrak{w}_\lambda|^2} \frac{1}{\varphi_\lambda^p} = \frac{|\mathfrak{w}_{\lambda p}|^2}{|\mathfrak{w}_{\lambda,p-1}|^2 |\mathfrak{w}_\lambda|^2 \varphi_\lambda^p}$$

$$\frac{\| \mathfrak{z}_{p+1} \|^2}{\| \mathfrak{z}_p \|^2} = \frac{|\mathfrak{w}_{\lambda,p-1}|^2 |\mathfrak{w}_{\lambda,p+1}|^2}{|\mathfrak{w}_{\lambda,p}|^4} \frac{1}{\varphi_\lambda} = \frac{H_p^\lambda}{\varphi_\lambda} = h_{up}$$

Hence

$$\frac{\| \mathfrak{z}_{p+1} \|^2}{\| \mathfrak{z}_p \|^2} = h_{up} \qquad \text{for } p = 0,1,\ldots,n-1.$$

<u>Therefore f is steady for B, if and only if a constant $C > 0$ exists such that the invariant flag ratio is bounded below.</u>

$$\frac{\| \mathfrak{z}_{p+1} \|}{\| \mathfrak{z}_p \|} \geqq C > 0 \text{ on } M^*.$$

3. Interpretation

<u>Steadiness of f for B means that the length of the pth torsion of f in the direction B is bounded below by a positive constant.</u>

For each $\lambda \in \Lambda$, define $\mathfrak{z}_{\lambda p}: U_\lambda \cap M^* \to V - \{0\}$ as in the 2. interpretation. Then

$$\mathfrak{y}_{\lambda p} = \frac{\mathfrak{z}_{\lambda p}}{|\mathfrak{z}_{\lambda p}|} \quad : U_\lambda \cap M^* \to V\langle 1\rangle$$

is a map of class C^∞ into the unit sphere in V. For each

$z \in U_\lambda \cap M^*$ an orthonormal base $\mathfrak{y}_{\lambda 0}(z), \ldots, \mathfrak{y}_{\lambda n}(z)$ of V is given such that

$$\mathfrak{y}_{\lambda 0} \wedge \cdots \wedge \mathfrak{y}_{\lambda p} = \frac{\mathfrak{w}_{\lambda p}}{|\mathfrak{w}_{\lambda p}|} \qquad p = 0, 1, \ldots, n$$

Then

$$d\,\mathfrak{y}_{\lambda p} = \sum_{q=0}^{n} \theta_{pq}^{\lambda}\, \mathfrak{y}_{\lambda q}$$

where θ_{pq}^{λ} is a form of degree 1 and class C^∞ on $U_\lambda \cap M_0$ defined by

$$\theta_{pq}^{\lambda} = (d\,\mathfrak{y}_{\lambda p} \mid \mathfrak{y}_{\lambda q})$$

Here the hermitian product on V is extended to differential forms in an obvious way. Because $(\mathfrak{y}_{\lambda p} \mid \mathfrak{y}_{\lambda q}) = \delta_{pq}$ on $U_\lambda \cap M^*$, this implies

$$0 = d(\mathfrak{y}_{\lambda p} \mid \mathfrak{y}_{\lambda q}) = (d\,\mathfrak{y}_{\lambda p} \mid \mathfrak{y}_{\lambda q}) + (\mathfrak{y}_{\lambda p} \mid d\,\mathfrak{y}_{\lambda q})$$

$$= \theta_{pq}^{\lambda} + \overline{\theta}_{qp}^{\lambda}$$

Therefore

$$\theta_{pq}^{\lambda} + \overline{\theta}_{qp}^{\lambda} = 0$$

If in doubt, the exterior product on V shall be denoted by $\wedge$ and the exterior product between differential forms shall be denoted by $\dot{\wedge}$. If no confusion will arise, both will be denoted by $\wedge$. Then

$$\mathfrak{w}_{\lambda p} = |\mathfrak{w}_{\lambda p}|\, \mathfrak{y}_{\lambda 0} \wedge \cdots \wedge \mathfrak{y}_{\lambda p}$$

$$\mathfrak{w}_{\lambda p-1} \wedge \mathfrak{w}_{\lambda}^{(p+1)} \zeta_{\lambda} = d\,\mathfrak{w}_{\lambda p} \wedge B$$

$$= (d|\mathfrak{w}_{\lambda p}| \,\dot{\wedge}\, B)\ \mathfrak{y}_{\lambda 0} \wedge \cdots \wedge \mathfrak{y}_{\lambda p}$$

$$+ |\mathfrak{w}_{\lambda p}| \sum_{q=0}^{p} \mathfrak{y}_{\lambda 0} \wedge \cdots \wedge \mathfrak{y}_{\lambda, q-1} \wedge (d\mathfrak{y}_{\lambda q} \,\dot{\wedge}\, B) \wedge \mathfrak{y}_{\lambda q+1} \wedge \cdots \wedge \mathfrak{y}_{\lambda p}$$

$$= [(d|\mathfrak{w}_{\lambda p}| \,\dot{\wedge}\, B) + |\mathfrak{w}_{\lambda p}| (\sum_{q=0}^{p} \theta_{pq}^{\lambda} \,\dot{\wedge}\, B)]\ \mathfrak{y}_{\lambda 0} \wedge \cdots \wedge \mathfrak{y}_{\lambda p}$$

$$+ |\mathfrak{w}_{\lambda p}| \sum_{q=0}^{p} \sum_{k=p+1}^{n} \mathfrak{y}_{\lambda 0} \wedge \cdots \wedge \mathfrak{y}_{\lambda, q-1} \wedge \mathfrak{y}_{\lambda k} \wedge \mathfrak{y}_{\lambda q+1} \wedge \cdots \wedge \mathfrak{y}_{\lambda p}$$

$$(\theta_{qk}^{\lambda} \,\dot{\wedge}\, B)$$

If $0 \leqq r \leqq p-1$, then $\mathfrak{y}_{\lambda r} \wedge \mathfrak{w}_{\lambda p-1} = 0$. Hence

$$0 = |\mathfrak{w}_{\lambda p}| (-1)^{p-1} \sum_{k=p+1}^{n} \mathfrak{y}_{\lambda 0} \wedge \cdots \wedge \mathfrak{y}_{\lambda p} \wedge \mathfrak{y}_{\lambda k} (\theta_{rk}^{\lambda} \,\dot{\wedge}\, B)$$

Hence $\Theta_{rk}^{\lambda} \,\dot{\wedge}\, B = 0$ if $0 \leqq r \leqq p-1$ and $n \geqq k \geqq p+1$ and $0 \leqq p \leqq n$. Therefore

$$\theta_{pq}^{\lambda} \wedge B = 0 \qquad \text{if } p+2 \leqq q \leqq n$$

$$\theta_{pq}^{\lambda} \wedge \overline{B} = 0 \qquad \text{if } 0 \leqq q \leqq p-2$$

Read $\theta^{\lambda}_{p,q}\,\mathfrak{y}_{\lambda q} \equiv 0$ if $q < 0$ or $q > n$. The following Frenet Formulas relative to B are obtained.

$$d\,\mathfrak{y}_{\lambda p} \wedge B \wedge \overline{B} = (\theta^{\lambda}_{pp-1}\,\mathfrak{y}_{\lambda p-1} + \theta^{\lambda}_{pp}\,\mathfrak{y}_{\lambda p} + \theta^{\lambda}_{pp+1}\,\mathfrak{y}_{\lambda p+1}) \wedge B \wedge \overline{B}$$

Here $\theta^{\lambda}_{p,p+1} \wedge B$ is called the pth torsion of the map f in direction B. Now

$$\mathfrak{y}_{\lambda p} \wedge \mathfrak{w}_{\lambda p-1} \wedge \mathfrak{w}^{(p+1)}_{\lambda}\,\zeta_{\lambda} = |\mathfrak{w}_{\lambda p}|(-1)^{p-1}\mathfrak{y}_{\lambda 0} \wedge \cdots \wedge \mathfrak{y}_{\lambda p+1}\,\theta^{\lambda}_{pp+1} \wedge B$$

$$|\mathfrak{w}_{\lambda p-1}|\,\mathfrak{y}_{\lambda 0} \wedge \cdots \wedge \mathfrak{y}_{\lambda p} \wedge \mathfrak{w}^{(p+1)}_{\lambda}\,\zeta_{\lambda} = |\mathfrak{w}_{\lambda p}|\,|\mathfrak{w}_{\lambda p+1}|^{-1}\mathfrak{w}_{\lambda p+1}\,\theta^{\lambda}_{pp+1} \wedge B$$

$$\frac{|\mathfrak{w}_{\lambda p-1}|}{|\mathfrak{w}_{\lambda p}|^{2}}\,\mathfrak{w}_{\lambda,p+1}\,\zeta_{\lambda} = \frac{|\mathfrak{w}_{\lambda p}|}{|\mathfrak{w}_{\lambda,p+1}|}\,\mathfrak{w}_{\lambda p+1}\,\theta^{\lambda}_{pp+1} \wedge B$$

Therefore

$$\theta^{\lambda}_{p\ p+1} \wedge B = \frac{|\mathfrak{w}_{\lambda p-1}|\ |\mathfrak{w}_{\lambda p+1}|}{|\mathfrak{w}_{\lambda p}|^{2}}\,\zeta_{\lambda}$$

Especially it follows that $\theta^{\lambda}_{pp+1} \wedge B$ has bidegree $(m,0)$, and can be considered as a section of class C^{∞} of K over $U_{\lambda} \cap M^{*}$. Therefore

$$\kappa(\theta^{\lambda}_{p,p+1} \wedge B) = \frac{|\mathfrak{w}_{\lambda p-1}|^{2}|\mathfrak{w}_{\lambda p+1}|^{2}}{|\mathfrak{w}_{\lambda p}|^{4}}\,\kappa(\zeta_{\lambda})$$

$$= \frac{H_{p\lambda}}{\varphi_{\lambda}} = h_{up}$$

is defined and independent of λ, namely

$$\kappa\,(\theta^{\lambda}_{p,p+1} \wedge B) = h_{up}$$

on $U_\lambda \cap M^*$. Therefore h_{up} is the square of the length of the pth torsion of the map f in direction B. The map f is steady for B if and only if this length is bounded below by a positive constant.

It is instructive to consider the Frenet formulas a step further. Obviously

$$H_p = i_0\, \theta^{\lambda}_{pp+1} \wedge \overline{\theta}^{\lambda}_{pp+1} \wedge B \wedge \overline{B}$$

The differential form θ^{λ}_{pq} has degree 1. Hence

$$\theta^{\lambda}_{pq} = \theta^{\lambda 0}_{pq} + \theta^{\lambda 1}_{pq}$$

with

$$\text{bideg}\ \theta^{\lambda 0}_{pq} = (1,0) \qquad \text{bideg}\ \theta^{\lambda 1}_{pq} = (0,1)$$

Split the formula for $\mathfrak{w}_{\lambda p-1} \wedge \mathfrak{w}^{(p+1)}_{\lambda}\zeta_\lambda$ by bidegree. Then

$$0 = (\overline{\partial}|\mathfrak{w}_{\lambda p}| \wedge B + |\mathfrak{w}_{\lambda p}| \sum_{q=0}^{p} \theta^{\lambda 1}_{qq} \wedge B)\,\mathfrak{y}_{\lambda 0} \wedge \cdots \wedge \mathfrak{y}_{\lambda p}$$

$$\overline{\partial}|\mathfrak{w}_{\lambda p}| \wedge B = -|\mathfrak{w}_{\lambda p}| \sum_{q=0}^{p} \theta^{\lambda 1}_{qq} \wedge B$$

Now

$$\theta^{\lambda 0}_{qq} + \theta^{\lambda 1}_{qq} = \theta^{\lambda}_{qq} = -\overline{\theta}^{\lambda}_{qq} = -\overline{\theta}^{\lambda 1}_{qq} - \overline{\theta}^{\lambda 0}_{qq}$$

implies $\theta_{qq}^{\lambda 0} = -\overline{\theta}_{qq}^{\lambda 1}$. Hence

$$\partial \,|\mathfrak{w}_{\lambda p}| \wedge \overline{B} = |\mathfrak{w}_{\lambda p}| \sum_{q=0}^{p} \theta_{qq}^{\lambda 0} \wedge \overline{B}$$

Therefore

$$d^{\perp} |\mathfrak{w}_{\lambda p}| \wedge \overline{B} = i\,|\mathfrak{w}_{\lambda p}| \sum_{q=0}^{p} \theta_{qq}^{\lambda} \wedge B \wedge \overline{B}$$

Hence

$$d^{\perp} \log |\mathfrak{w}_{\lambda p}| \wedge B \wedge \overline{B} = i \sum_{q=0}^{p} \theta_{qq}^{\lambda} \wedge B \wedge \overline{B}$$

In the case $m = 1$ these Frenet formulas were obtained by Chern [8] section 3. They are extended here to the case $m > 1$.

4. Interpretation

<u>Invariant holomorphic hermitian line bundles $Q_0, \ldots, Q_n$ are associated to f and B such that Q_p is differentiably isomorphic to $Q_{p+1} \otimes K$. The hermitian metric along the fibers of the bundles Q_p define hermitian metrics along K.</u> The <u>map f is steady for B if and only if these induced metrics along the fibers of K are majorized by the hermitian metric κ defined by φ_u^m.</u>

Take $z \in M^*$. Take any $\lambda \in \Lambda$ with $z \in U_\lambda$. Then

$$L_{pz} = \{ \mathfrak{z} \in V \mid \mathfrak{z} \wedge \mathfrak{w}_{p\lambda}(z) = 0 \} = E(\mathfrak{w}_{p\lambda}(z))$$

is a linear subspace of dimension p+1 of V which is independent of the choice of $\lambda \in \Lambda$. The disjoint union L_p of all L_{pz} with $z \in M^*$ is a holomorphic vector bundle over M^* with fiber dimension p+1. Then

$$L_0 \subset L_1 \subset \ldots \subset L_n = V_{M^*}$$

is a sequence of subbundles of the trivial bundle V_{M^*}. This sequence shall be called the <u>banner of f for B</u>. The hermitian metric on V defines a hermitian metric along the fibers of V_{M^*} which restricts to a hermitian fiber along the fiber of L_p for $p = 0,1,\ldots,n$. Observe that $L_0 = L|M^*$ as hermitian vector bundle. Define the quotient bundle Q_p for $p = 0,1,\ldots,n$ by

$$Q_0 = L_0 \qquad Q_p = L_p/L_{p-1} \quad p = 1,\ldots,n$$

over M^*. These holomorphic line bundles $Q_0,\ldots,Q_n$ are called the invariant line bundles of f for B. Let $\iota_p\colon L_{p-1} \to L_p$ be the inclusion and $\rho_p\colon L_p \to Q_p$ the residual map. The exact sequence

$$0 \to L_{p-1} \xrightarrow{\iota_p} L_p \xrightarrow{\rho_p} Q_p \to 0$$

splits differentiably $L_p = L_{p-1} \oplus L_{p-1}^{\perp}$, where $L_{p-1}^{\perp}$ is a differentiable complex line bundle such that

$$L_{p-1\,z}^{\perp} = \{\mathfrak{z} \in L_{pz} \mid (\mathfrak{z}\,|\,\mathfrak{y}) = 0 \text{ for all } \mathfrak{y} \in L_{p-1,z}\}$$

if $z \in M^*$. The restriction $\rho_p\colon L_{p-1}^{\perp} \to Q_p$ is a differentiable line bundle isomorphism. The hermitian metric along the fibers of L_p restricts to a hermitian metric along the fibers of $L_{p-1}^{\perp}$ and the isomorphism $\rho_p | L_{p-1}^{\perp}$ carries this hermitian metric over to a hermitian metric along the fibers of Q_p. If $w \in Q_{pz}$, denote by $|w|_p$ the length of w in this induced hermitian metric.

Now a differentiable isomorphism

$$\sigma_p\colon Q_{p-1} \to Q_p \otimes K$$

shall be defined. Take $z \in M^*$. Take any $\lambda \in \Lambda$ with $z \in U_\lambda$. Take $w \in Q_{p-1,z}$. Then $\mathfrak{w} \in L_{p-1,z}$ exists such that $\rho_{p-1}(\mathfrak{w}) = w$. Then

$$\mathfrak{w} = w_0^\lambda \ \mathfrak{r}_{0\lambda}(z) + \dots + w_{p-1}^\lambda \ \mathfrak{r}_{p-1\lambda}(z)$$

Define

$$\sigma_p(w) = w_{p-1}^\lambda \ \rho_p(\ \mathfrak{r}_{p\lambda}(z)) \otimes \zeta_\lambda(z)$$

Obviously, $\sigma_p(w) \in Q_{pz} \otimes K_z$. It has to be shown that σ_p is well defined. Let $z \in U_\mu$ with $\mu \in \Lambda$. Take $\rho_{p-1}(\tilde{\mathfrak{w}}) = w$. Then

$$\tilde{\mathfrak{w}} = \tilde{w}_0^\mu \ \mathfrak{r}_{\mu 0}(z) + \dots + \tilde{w}_{p-1}^\mu \ \mathfrak{r}_{\mu p-1}(z)$$

$$= \sum_{q=0}^{p-1} w_q^\lambda \ \mathfrak{r}_{\lambda q}(z) = \sum_{q=0}^{p-1} w_q^\lambda \ g_{\lambda\mu}(z)\Delta_{\mu\lambda}^q(z) \ \mathfrak{r}_{\mu q}(z)$$

Now, $\rho_{p-1}(\mathfrak{E}_{\mu,p-1}(z)) \neq 0 \neq \rho_{p-1}(\mathfrak{E}_{\lambda,p-1}(z))$. Hence

$$w = \rho_{p-1}(\tilde{\mathfrak{w}}) = \tilde{w}^{\mu}_{p-1}\,\rho_{p-1}(\mathfrak{E}_{\mu,p-1}(z))$$

$$w = \rho_{p-1}(\mathfrak{w}) = w^{\lambda}_{p-1}\,\rho_{p-1}(\mathfrak{E}_{\lambda,p-1}(z))$$

$$= w^{\lambda}_{p-1}\, g_{\lambda\mu}\, \Delta^{p-1}_{\mu\lambda}\, \rho_{p-1}\,(\mathfrak{E}_{\mu,p-1}(z))$$

Hence

$$\tilde{w}^{\mu}_{p-1} = w^{\lambda}_{p-1}\, g_{\lambda\mu}\, \Delta^{p-1}_{\mu\lambda}$$

Therefore

$$\tilde{w}^{\mu}_{p-1}\,\rho_p(\mathfrak{E}_{\mu p}(z)) \otimes \zeta_\mu(z) =$$

$$= w^{\lambda}_{p-1}\;\; \lambda_\mu(z)(\Delta_{\mu\lambda}(z))^{p-1}\,\rho_p(\mathfrak{E}_{\mu p}(z)) \otimes \Delta_{\mu\lambda}(z)\zeta_\lambda(z)$$

$$= w^{\lambda}_{p-1}\,\rho_p(g_{\lambda\mu}(z)(\Delta_{\mu\lambda}(z))^p\,\mathfrak{E}_{\mu p}(z)) \otimes \zeta_\lambda(z)$$

$$= w^{\lambda}_{p-1}\,\rho_p(\mathfrak{E}_{\lambda p}(z)) \otimes \zeta_\lambda(z)$$

Therefore σ_p is well defined. Obviously, σ_p is a differentiable vector bundle homomorphism. If $\sigma_p(w) = 0$ then $w^{\lambda}_{p-1} = 0$. Hence $\mathfrak{w} \in L_{p-2,z}$ and $w = \rho_{p-1}(\mathfrak{w}) = 0$. Hence σ_p is a line bundle isomorphism of class C^∞. Then one and only one hermitian metric κ_p along the fibers of $K|M^*$ exist such that

$$|\sigma^{-1}(u \otimes v)|^2_{p-1} = |u|^2_p\, \kappa_p(v)$$

if $u \in Q_{pz}$ and $v \in K_z$ and $z \in M^*$.

If n_1 and n_2 are two hermitian metric along the fibers of a complex line bundle on M^*, then one and only function $\frac{n_1}{n_2} : M^* \to \mathbb{R}$ is defined by

$$\frac{n_1}{n_2}(z) = \frac{n_1(e)}{n_2(e)} \quad \text{for all } 0 \neq e \in E_z$$

The two hermitian metrics κ and κ_{p+1} along the fibers of $K|M^*$ define the quotient

$$q_p = \frac{\kappa}{\kappa_{p+1}} : M^* \to \mathbb{R}$$

for $p = 0,1,\ldots,n-1$. These quotients shall be computed.

Take $z \in M_0$. Take $w \in Q_{pz}$. Take $\mathfrak{w} \in L_{pz}$ with $\rho_p(\mathfrak{w}) = w$. Then

$$\mathfrak{w} = w_0^\lambda \ \mathfrak{E}_{\lambda 0}(z) + \ldots + w_p^\lambda \ \mathfrak{E}_{\lambda p}(z)$$

Now $\mathfrak{E}_{\lambda 0}(z),\ldots, \mathfrak{E}_{\lambda p-1}(z)$ span $L_{p-1,z}$ and $\mathfrak{E}_{\lambda p}(z)$ spans $L^{\perp}_{p-1,z}$. Hence $|w|_p = |w_p^\lambda| \ |\mathfrak{E}_{\lambda p}(z)|$. Also

$$\sigma_{p+1}(w) = w_p^\lambda \ \rho_p(\mathfrak{E}_{\lambda,p+1}(z)) \otimes \zeta_\lambda(z)$$

$$|w_p^\lambda|^2 \ |\mathfrak{E}_{\lambda p}(z)|^2 = |w|_p^2 = |w_p^\lambda|^2 \ |\mathfrak{E}_{\lambda p+1}(z)|^2 \ \kappa_{p+1}(\zeta_\lambda)$$

or

$$\kappa_{p+1}(\zeta_\lambda) = \frac{|\mathfrak{E}_{\lambda p}|^2}{|\mathfrak{E}_{\lambda p+1}|^2} = \frac{|\mathfrak{w}_{\lambda p}|^2}{|\mathfrak{w}_{\lambda,p-1}|^2} \ \frac{|\mathfrak{w}_{\lambda p}|^2}{|\mathfrak{w}_{\lambda,p+1}|^2} = \frac{1}{H_p^\lambda}$$

Therefore

$$q_p = \frac{\kappa}{\kappa_{p+1}} = \frac{\kappa(\zeta_\lambda)}{\kappa_{p+1}(\zeta_\lambda)} = \frac{H_p^\lambda}{\varphi_\lambda} = h_{up}.$$

Hence

$$q_p = \frac{\kappa}{\kappa_{p+1}} = h_{up}$$

gives another interpretation of h_{up} on M^*.

The map f is steady for B if and only if there exists a constant $C > 0$ such that $\kappa \geqq C\,\kappa_{p+1}$ on M^* for $p = 0,1,\ldots,n-1$, that is if κ majorizes κ_{p+1}.

Now, maps with sufficient growth shall be considered. Assume (B1) - (B9). A meromorphic map f is said to grow sufficiently if and only if there exists a number ε with $0 \leqq \varepsilon < \frac{1}{2}$, constants C_1, C_2, C_3, C_4 and a number $r_1 \geqq r_0$ such that for all $r \geqq r_1$ the following estimates hold

$$\Phi(r) \leqq C_1\, T_f(r)^\varepsilon \tag{13.1}$$

$$\log^+ Z_u(r) \leqq C_2\, T_f(r)^\varepsilon \tag{13.2}$$

$$\log^+ \mathrm{Ric}_u^+(r) \leqq C_3\, T_f(r)^\varepsilon \tag{13.3}$$

$$\log^+ Q_u(r) \leqq C_4\, T_f(r)^\varepsilon \tag{13.4}$$

The following additional general assumption will be made.

(B11) The meromorphic map f grows sufficiently.

Observe that assumption (B11) does not depend on the choice of B.

Let f be adapted to $a \in \mathbb{P}(V^*)$. Then

$$T_f(r) = N_f(r,a) + m_f(r,a) - D_f(r,a)$$

for all $r \in E_\tau$. Here T_f, N_f, D_f are continuous functions of r for all $r \geqq 0$. Hence $m_f(r,a)$ continuous to a continuous function for all $r \geqq 0$. The <u>defect</u> of f for $a \in \mathbb{P}(V^*)$ is defined by

$$\delta_f(a) = \liminf_{r \to +\infty} \frac{m_f(r,a)}{T_f(r)} \geqq 0$$

The defect does not depend on B. The existence of the deficit does not allow a trivial upper estimate of $\delta_f(a)$. However the following results can be obtained.

<u>Lemma 13.8</u>. Assume (B1) - (B11). Take $a \in \mathbb{P}(V^*)$. Assume that $f^{-1}(\ddot{E}[a]) = \emptyset$. Then $\delta_f(a) \geqq 1$.

Proof. The First Main Theorem implies

$$\delta_f(a) = \liminf_{r \to \infty} \frac{m_f(r,a)}{T_f(r)} = \liminf_{r \to \infty} \left(1 + \frac{D_f(r,a)}{T_f(r)}\right) \geqq 1$$

q.e.d.

<u>Theorem 13.9</u>. Assume (B1) - (B11). Take $a \in \mathbb{P}(V^*)$. Then

$$0 \leqq \delta_f(a) \leqq 1$$

Proof. There exist $0 \leqq \varepsilon < \frac{1}{2}$ and $C_\lambda \geqq 0$ for $\lambda = 1,2,3,4$ and $r_1 > r_0$ such that the estimates (13.1) - (13.4) hold. Then

$$\log^+ T_f(r) \leqq \frac{1}{\varepsilon} T_f(r)^\varepsilon \qquad \text{for } r_1 > r_0$$

Theorem 13.4 implies

$$D_f(r,a) \leqq C\, c_1 [\tfrac{1}{\varepsilon} + c_2 + c_4 + c_1 + c_3]\, T_f^{2\varepsilon}(r) = c_5\, T_f^{2\varepsilon}(r)$$

for all $r > r_1$ except a set of finite measure. Here c_5 is a constant. Because $1 - 2\varepsilon > 0$, this implies

$$0 \leqq \delta_f(a) = \liminf_{r \to +\infty} \frac{m_f(r,a)}{T_f(r)} \leqq \liminf_{r \to \infty} (1 + \frac{D_f(r,a)}{T_f(r)} = 1$$

q.e.d.

Because of the existence of the exceptional sets the classical identity

$$\delta_f(a) = 1 - \limsup_{r \to \infty} \frac{N_f(r,a)}{T_f(r)}$$

may be false even under the assumptions of Theorem 13.9.

The meromorphic map $f: M \longrightarrow \mathbb{P}(V)$ is said to grow slowly if and only if constants $C_0 > 0$ and θ and r_1 with $1 < \theta \leqq 2 < r_1$ exist such that

$$T_f(\theta r) \leqq C_0\, T_f(r) \qquad \text{for all } r \geqq r_1 \tag{13.5}$$

The following general assumption shall be made.

(B12) The meromorphic map f grows slowly.

Theorem 13.10. Assume (B1) - (B12). Then

$$\frac{D_f(r,a)}{T_f(r)} \Longrightarrow 0 \qquad \text{for } r \to \infty$$

uniformly for all $a \in \mathbb{P}(V^*)$. Moreover

$$\delta_f(a) = 1 - \limsup_{r \to \infty} \frac{N_f(r,a)}{T_f(r)}$$

Proof. Let θ, r_1, c_0, $c_1, \ldots, c_4$ be constants such that (13.1) - (13.5) hold. For $a \in \mathbb{P}(V^*)$, Theorem 13.5 implies

$$D_f(r,a) \leqq 2^{2n+1} c_1 \left[\frac{1}{\varepsilon} + c' \log^+ c_1 + (m-1)c_2 + c_4 + c_3\right](T_f(r) T_f(\theta r))^{\varepsilon}$$

$$+ 2^{2n+1} c_1 \left[\frac{\theta}{\theta-1} + c''\right] T_f(r)^{\varepsilon} + 2^n$$

for all $r \geqq r_1$. Here c' and c'' are constants independent of r and $a \in \mathbb{P}(V^*)$. Because T_f is not bounded and because $T_f(\theta r) \leqq c_0 T_f(r)$ for $r \geqq r_1$, constants $c_5 > 0$ and $r_2 > r_1$ exist which do not depend on r and $a \in \mathbb{P}(V^*)$ such that

$$D_f(r,a) \leqq c_5 T_f(r)^{2\varepsilon}$$

for all $r \geqq r_2$ and all $a \in \mathbb{P}(V^*)$. Hence

$$\frac{D_f(r,a)}{T_f(r)} \leqq \frac{c_5}{T_f(r)^{1-2\varepsilon}} \qquad \text{if } r \geqq r_2$$

Now $0 \leqq 2\varepsilon < 1$ implies

$$\frac{D_f(r,a)}{T_f(r)} \Longrightarrow 0 \qquad \text{for } r \to +\infty$$

uniformly on $\mathbb{P}(V^*)$. Also

$$\delta_f(a) = \liminf_{r \to \infty} \frac{m_f(r,a)}{T_f(r)} = \liminf_{r \to \infty} \frac{m_f(r,a) - D_f(r,a)}{T_f(r)}$$

$$= 1 - \limsup_{r \to \infty} \frac{N_f(r,a)}{T_f(r)}$$

by the first main theorem, q.e.d.

Under the assumption (B1) - (B12) it would be easy to obtain a theorem of Casorati-Weierstrass type. However such a theorem was already obtained in [41] Theorem 4.5 respectively [42] Theorem 9.8 (see also page 134) and [39] Theorem 5.12 for s = 1.

XIV. THE BEZOUT ESTIMATES

The theory shall be applied to solve a Bezout problem posed by Phillip Griffiths in the case where a close, smooth, connected, complex submanifold of a complex vector space is intersected by a linear subspace of codimension 1.

A new set of <u>general assumptions</u> will be made.

(C1) Let V be a hermitian vector space of dimension n+1 with $n > 0$. Define $\tau_0 \colon V \to \mathbb{R}$ by $\tau_0(\mathfrak{z}) = |\mathfrak{z}|^2$ if $\mathfrak{z} \in V$.

Recall the notations of section 2. Observe that the hermitian metric is denoted here by τ_0 not by τ as in section 2. Observe that A[r], A(r), A<r> are the intersections of $A \subseteq V$ with the closed ball, respectively open ball, respectively the sphere all of radius r and center 0 in V. Observe that

$$\upsilon = \frac{1}{4\pi} dd^c \tau_0 \qquad \text{on } V$$

$$\omega = \frac{1}{4\pi} dd^c \log \tau_0 \quad \text{on } V - \{0\}$$

are defined.

Let A be an analytic subset of pure dimension p of V. If $p = 0$, define the counting function by

$$n_A(r) = \#A(r) \qquad r \geqq 0$$

If $p > 0$, define the counting function by

$$n_A(r) = \frac{1}{r^{2p}} \int_{A(r)} \upsilon^p = \int_{A(r)} \omega^p + n_A(0) \qquad \text{if } r > 0$$

Here $n_A(0) = \lim_{r \to 0} n_A(r)$ is the Lelong number (see Lelong [24]) which is an integer by Thie [46]. The function n_A increases and is continuous for $p > 0$. If $p > 0$, the set A[r] has measure zero on A. Hence the integration over A(r) equals the integration over A[r]. If $0 < s < r$ define the valence function by

$$N_A(r,s) = \int_s^r n_A(t) \frac{dt}{t}$$

The function N_A is continuous and increases with r and decreases with s. If $0 \notin A$, define

$$N_A(r) = N_A(r,0) = \int_0^r n_A(t)\,\frac{dt}{t}$$

Now, the following additional <u>general assumptions</u> shall be made.

(C2) Let M be a connected, closed, smooth, complex submanifold of dimension $m > 0$ of V with $M[1] = \emptyset$. Let $\iota : M \to V$ be the inclusion map.

(C3) Assume that M is not contained in any proper linear subspace of V.

(C2) is equivalent to the requirement that $M \not\subseteq E[a]$ for all $a \in \mathbb{P}(V^*)$. Also (C1) implies $M \neq V$, hence $m \leqq n$.

Let $\nu: M \to \mathbb{Z}$ be any divisor on M. Let $\gamma(\nu)$ be the support of ν. If $m = 1$, define the <u>counting function</u> by

$$n_\nu(r) = \sum_{|z|<r} \delta_\nu(z)$$

If $m > 1$, define the <u>counting function</u> by

$$n_\nu(r) = r^{2-2m} \int_{\gamma(\nu)(r)} \nu\upsilon^{m-1} = \int_{\gamma(\nu)(r)} \nu\,\omega^{m-1}$$

In this case n_ν is continuous. If $\nu \geqq 0$, then n_ν increases in both cases. The <u>valence function</u> of ν is defined for $0 \leqq s < r$ by

$$N_\nu(r,s) = \int_s^r n_\nu(t)\,\frac{dt}{t}$$

$$N_\nu(r) = N_\nu(r,0) = \int_0^r n_\nu(t)\,\frac{dt}{t}$$

The function N_ν is continuous. It increases with r and decreases with s if $\nu \geqq 0$.

If $a \in \mathbb{P}(V^*)$, then $M_a = E[a] \cap M$ is thin analytic in M according to (C2) and (C3). Take any $\alpha \in \mathbb{P}^{-1}(a)$. Then $\alpha \circ \mathfrak{w} \not\equiv 0$ is a holomorphic function on M. Let $\nu = \delta_M^a = \delta_{\alpha \circ \mathfrak{w}}$ be its zero divisor. Define $n_M(r,a) = n_\nu(r)$ as the counting function of M for a and $N_M(r,s;a) = N_\nu(r,s)$ as the valence function of M for a.

The Bezout problem is to estimate $N_M(r,a)$ in terms of $N_M(r)$ and perhaps other invariants of M. For this the assumptions (Bx) will be satisfied by assumption (Cy) on M.

A holomorphic map

$$f = P \circ \mathfrak{w} \ : M \to \mathbb{P}(V)$$

is defined. Here $\mathfrak{w}$ is a simple global representation of f. The map f is generic by (C2). Observe $\delta_M^a = \delta_f^a$ and

$$f^{-1}(\ddot{E}[a]) = M_a = M \cap E[a] = \gamma(\delta_M^a) = \gamma(\delta_f^a)$$

for all $a \in \mathbb{P}(V^*)$.

Define

$$\tau = \frac{1}{2}\log \tau_0 \circ \mathfrak{w} \ : M \to \mathbb{R}$$

Then τ is a positive function of class C^∞ on M. The set

$$\Gamma_r = \tau^{-1}(r) = M\langle e^r\rangle \qquad \text{if } r > 0$$

is real analytic and does not contain any positive dimensional analytic subset of M. Also Γ_r has finite $\mathfrak{H}^{2m-1}$-measure. Observe

$$G_r = \{z \in M \mid \tau(z) < r\} = M(e^r) \quad \text{if } r > 0$$

Here $\Gamma_r - dG_r$ has zero $\mathfrak{H}^{2m-1}$-measure on M. The map τ is an exhaustion on M with $E_\tau = \mathbb{R}[0,+\infty)$. Also

$$M(r) = G_{\log^+ r} \qquad M\langle r\rangle = \Gamma_{\log^+ r}$$

The associated differential form φ is defined by

$$\varphi = \frac{1}{2\pi} dd^\perp\tau = \mathfrak{v}^*(\omega) = f^*(\ddot{\omega}) \geqq 0$$

<u>Lemma 14.1</u>. Assume (C1) - (C3). The holomorphic map f is light. Let S be the set of all $x \in M$ where the Jacobian matrix of f at x has a rank smaller than m. Then S is thin analytic in M. Moreover $M - \text{pos}\varphi \subseteqq S$. Especially, τ is a pseudo-convex exhaustion of M.

Proof. Take $y \in \mathbb{P}(V)$. Take any $\mathfrak{y} \in \mathbb{P}^{-1}(y)$. Then $\mathfrak{y} \neq 0$. Define $\mathbb{C}\mathfrak{y} = \{z\mathfrak{y} \mid z \in \mathbb{C}\}$

$$f^{-1}(y) = M \cap \mathbb{P}^{-1}(y) = M \cap \mathbb{C}\mathfrak{y}$$

is analytic in $\mathbb{C}\mathfrak{y}$. If dim $f^{-1}(y) > 0$, then $f^{-1}(y) = \mathbb{C}\mathfrak{y}$. Hence $0 \in f^{-1}(y) \subseteq M$ which is false. Hence dim $f^{-1}(y) \leqq 0$ for all $y \in \mathbb{P}(V)$. The map f is light. Hence S is thin analytic on M. If $x \in M-S$, then $f: M \to \mathbb{P}(V)$ is an immersion in a neighborhood of x. Hence $\varphi(x) = f^*(\ddot{\omega})(x) > 0$. Therefore $M - \text{pos}\varphi \subseteq S$;

q.e.d.

Therefore the assumption (B1) - (B4) are satisfied. Define

$$r_1 = e^{r_0} > 1$$

Hence $M(r) = \emptyset$ if $r \quad r_1$ and $M(r) \neq \emptyset$ if $r > r_1$. Also

$$A_f(r) = \int_{G_r} f^*(\ddot{\omega}) \wedge \varphi^{m-1} = \int_{G_r} \varphi^m = \int_{M[e^r]} \omega^m = n_M(e^r)$$

$$\Phi(r) = \int_{G_r} \varphi^m = n_M(e^r)$$

if $r \geqq 0$. Hence

$$n_M(r) = A_f(\log^+ r) = \Phi(\log^+ r)$$

for all $r \geqq 0$. If $0 \leqq s < r$ then

$$T_f(r,s) = \int_s^r A_f(t)dt = \int_{e^s}^{e^r} n_M(t)\frac{dt}{t} = N_M(e^r, e^s)$$

$$\Phi_1(r) = \int_0^r \Phi(t)dt = \int_0^r A_f(t)dt = N_M(e^r, 1) = N_M(r)$$

Hence

$$N_M(r,s) = T_f(\log^+ r, \log^+ s) \qquad \text{if } 0 \leqq s < r$$

$$N_M(r) = \Phi_1(r) = T_f(\log^+ r) \qquad \text{if } 0 \leqq r$$

If $a \in \mathbb{P}(V^*)$, then

$$n_f(r,a) = \int_{\gamma_f(a) \cap G_r} \delta_f^a \, \varphi^{m-1} = \int_{M_a[e^r]} \delta_M^a \, \omega^{m-1} = n_M(e^r, a)$$

$$N_f(r,s;a) = \int_s^r n_f(t;a)dt = \int_{e^s}^{e^r} n_M(t;a)\frac{dt}{t} = N_f(e^r, e^s, a)$$

Hence

$$n_M(r;a) = n_f(\log^+ r; a)$$

$$N_M(r,s,a) = N_f(\log^+ r, \log^+ s; a)$$

$$N_M(r;a) = N_f(\log^+ r; a)$$

Define

$$\sigma = \frac{1}{4\pi} d^c \log \tau_0 \wedge \omega^{m-1} \qquad \text{on } V - \{0\}$$

$$\rho = \frac{1}{4\pi} d^c \tau_0 \wedge \upsilon^{m-1} \qquad \text{on } V$$

According to [45] p. 387-388

$$d\sigma = \omega^m \qquad d\rho = \upsilon^m \qquad \sigma = \tau^{-m}\rho$$

Let $j_r\colon M\langle r\rangle \to V$ be the inclusion map. Then

$$\sigma_r = j_r^*(\sigma) = \frac{1}{r^{2m}}\ j_r^*(\rho) \geqq 0$$

on $M\langle r\rangle$. Since $M(r)$ is a Stokes domain on M, Tung's Stokes Theorem implies

$$n_M(r) = r^{-2m} \int_{M[r]} \upsilon^m = r^{-2m} \int_{M\langle r\rangle} \rho =$$

$$= \int_{M\langle r\rangle} \sigma = \int_{M\langle r\rangle} \sigma_r$$

Hence

$$n_M(r) = \int_{M\langle r\rangle} \sigma_r$$

for all $r > 0$. If $a \in \mathbb{P}(V^*)$, define the <u>compensation function</u> by

$$m_M(r;a) = \int_{M\langle r\rangle} \log \frac{1}{\|f;a\|}\ \sigma_r = \frac{1}{2\pi} \int_{M\langle r\rangle} \log \frac{1}{\|f;a\|} d^c\tau \wedge \varphi^{m-1}$$

Therefore

$$m_M(r;a) = m_f(\log^+ r;a)$$

The <u>deficit</u> is defined similarly

$$D_M(r,a) = \int_{M(r)} \log \frac{1}{\|f;a\|}\ \omega^m = D_f\ (\log^+ r;a)$$

Consider the bump $(G_r, \emptyset, \psi_r)$ with $\psi_r = r-\tau$ on G_r and with $\chi = \varphi^{m-1}$. Then $\frac{1}{2\pi} dd^{\perp} \psi_{rs} \wedge \varphi^{m-1} = \varphi^m$ and $\frac{1}{2\pi} d^{\perp} \psi_r \wedge \chi = \sigma$ Therefore

$$\frac{1}{4\pi} \int_{G_r} dd^{\perp}\psi_r \wedge \varphi^{m-1} = \frac{1}{2} n_M(e^r)$$

$$\frac{1}{4\pi} \int_{\Gamma_r} d^{\perp}\psi_r \wedge \varphi^{m-1} = \frac{1}{2} \int_{M\langle e^r\rangle} \sigma_r = \frac{1}{2} n_M(e^r).$$

Hence Theorem 6.6 and Theorem 10.7 imply

Theorem 14.2. First Main Theorem.
Assume (C1) - (C3). Take $a \in \mathbb{P}(V^*)$. Then

$$N_M(r) = N_M(r;a) + m_M(r;a) - D_M(r;a)$$

$$N_M(r) = \int_{\mathbb{P}(V^*)} N_M(r;a)\, \ddot{\omega}^n(a)$$

$$\frac{1}{2} \sum_{\nu=1}^{n} \frac{1}{\nu} n_M(r) = \int_{\mathbb{P}(V^*)} m_M(r;a)\, \ddot{\omega}^n(a)$$

$$\frac{1}{2} \sum_{\nu=1}^{n} \frac{1}{\nu} n_M(r) = \int_{\mathbb{P}(V^*)} D_M(r;a)\, \ddot{\omega}^n(a)$$

for all $r \geqq 0$.

Of course, the First Main Theorem holds also for a meromorphic map $h: M \longrightarrow \mathbb{P}(W)$, where W is some hermitian vector space, say of dimension k+1 with $k \in \mathbb{N}$. Let $\tilde{\omega}$ be the Kaehler form of the Fubini Study metric on P(W). Define

$$\hat{A}_h(r) = \int_{M(r)} h^*(\tilde{\omega}) \wedge \omega^{m-1}$$

$$\hat{T}_h(r) = \int_0^r \hat{A}_h(t)\,\frac{dt}{t}$$

Then

$$\hat{A}_h(r) = A_h\,(\log^+ r)$$

$$\hat{T}_h(r) = T_h\,(\log^+ r)$$

Define u: $\mathbb{R}[0,+\infty) \to \mathbb{R}[1,+\infty)$ by

$$u(r) = e^{2r} \qquad u'(r) = 2e^{2r} \qquad u''(r) = 4e^{2r}$$

Hence $u \in \mathfrak{U}$. Observe

$$u(\log r) = r^2 \qquad u'(\log r) = 2r^2 \qquad u''(\log r) = 4r^2$$

if $r \geqq 1$.

$$u \circ \tau = e^{2\tau} = \tau_0 \circ \mathbf{10}$$

$$\varphi_u = \frac{1}{2\pi}\, dd^c\, u \circ \tau = \frac{1}{2\pi}\, dd^c\, \tau \circ \mathbf{10} = 2\ \mathbf{10}^*(\upsilon) > 0$$

Hence u is a strong convex change of scale for τ. Now, the following general assumption shall be made

(C4) Take $u \in \mathfrak{U}$ with $u(r) = e^{2r}$ as a strong convex change of scale of τ.

Proposition 14.3. Assume (C1) - (C4). Then

$$0 \leqq \rho_u \leqq \frac{1}{2m} \left(\frac{1}{2\tau_0 \circ \mathfrak{w}}\right)^m < 1$$

on M. Also $Q_u \equiv 0$.

Proof. By definition

$$\frac{1}{2\pi} d\tau \wedge d^c\tau \wedge \varphi^{m-1} = \rho_u \varphi_u^m$$

Observe

$$0 \leqq \tau_0^2\omega = \tau_0 \upsilon - \frac{1}{4\pi} d\tau_0 \wedge d^c\tau_0$$

Then

$$0 \leqq \tau_0^{2m} \omega^m = \tau_0^m \upsilon_0^m - \frac{m}{4\pi} \tau_0^{m-1} d\tau_0 \wedge d^c\tau_0 \wedge \upsilon^{m-1}$$

$$\frac{m}{4\pi} d^{\perp}\tau_0 \wedge d\tau_0 \wedge \omega^{m-1} = \frac{m}{4\pi} d\tau_0 \wedge d^c\tau_0 \wedge \upsilon^{m-1} \tau_0^{1-m}$$

$$\leqq \tau_0^{2-m} \upsilon^m$$

Hence

$$\rho_u \varphi_u^m = \frac{1}{2\pi} d\tau \wedge d^c\tau \wedge \varphi^{m-1}$$

$$= \mathfrak{w}^*\left(\frac{1}{2\pi} \tau_0^{-2} d\tau_0 \wedge d^c\tau_0 \wedge \omega^{m-1}\right)$$

$$\leqq \mathfrak{w}^*\left(\frac{1}{2m}\,\tau_0^{-m}\quad \upsilon^m\right)$$

$$= \frac{1}{2m}\,(2\tau_0 \circ \mathfrak{w})^{-m}\quad \varphi_u^m$$

Because $\tau_0 \circ \mathfrak{w} \geqq 1$ on M, this implies

$$0 \leqq \rho_u \leqq \frac{1}{2m}\left(\frac{1}{2\tau_0 \circ \mathfrak{w}}\right)^m < \frac{1}{m}\left(\frac{1}{2}\right)^{m+1} < 1$$

Therefore

$$Q_u(r) = \frac{1}{4\pi}\int_{G_r} \log^+\rho_u \; d\tau \wedge d^c\tau \wedge \varphi^{m-1} = 0$$

q.e.d.

The form φ_u^m defines a hermitian metric κ along the fibers of the canonical bundle K. Now Ric κ and $\mathrm{Ric}_u(r)$ shall be computed.

Take $z \in M$. Then there exists one and only one $\Delta(z) \in G_{m-1}(V) \subseteq \mathbb{P}(\underset{m}{\Lambda}V)$ such that $E(\Delta(z))$ is the linear subspace of V which is parallel to the tangent plane of M at z. A map

$$\Delta : M \to \mathbb{P}(\underset{m}{\Lambda}(V))$$

with $\Delta(M) \subseteq G_{m-1}(V)$ is defined which is called the <u>Gauss map.</u>

<u>Proposition 14.4</u>. Assume (C1) - (C4). Let κ be the hermitian metric along the fibers of the canonical bundle K which is defined by φ_u^m. Then the Gauss map $\Delta\colon M \to \mathbb{P}(\underset{m}{\Lambda}V)$ is holomorphic and

$$\text{Ric}\ \kappa = \Delta^*(\ddot{\omega}'_{m-1}) \geqq 0$$

$$\text{Ric}_u\ (\log^+ r) = \text{Ric}^+_u\ (\log^+ r) = \hat{T}_\Delta(r)$$

Proof. Let $\alpha \in \mathfrak{V}_M$ be a patch on M. Define

$$\xi = \mathfrak{w} \circ \alpha^{-1} : U'_\alpha \to V$$

Then ξ is a holomorphic embedding of U'_α into V with $\xi(U'_\alpha) = U_\alpha$. Let $w_1, \ldots, w_m$ be the coordinate functions on $\mathbb{C}^m \supseteq U'_\alpha$. Take $z \in M$. Define $\mathfrak{w} = \alpha(z)$. Then

$$E(\Delta(z)) = \{ \mathfrak{z} \in V \mid \mathfrak{z} \wedge \xi_{w_1}(\mathfrak{w}) \wedge \cdots \wedge \xi_{w_m}(\mathfrak{w}) = 0 \}$$

Therefore

$$\Delta | U_\alpha = \mathbb{P} \circ (\xi_{w_1} \circ \alpha) \wedge \cdots \wedge (\xi_{w_m} \circ \alpha)$$

is holomorphic on U_α. Now

$$(\alpha^{-1})^* \varphi_u = 2(\alpha^{-1})^* \mathfrak{w}^*(\upsilon) = 2\ \xi^*(\upsilon) = \frac{1}{2\pi}\, dd^c \tau \circ \xi$$

$$= \frac{i}{\pi} \sum_{\mu,\nu=1}^{m} (\xi_{w_\mu} \mid \xi_{w_\nu})\, dw_\mu \wedge d\bar{w}_\nu$$

$$(\alpha^{-1})^* \varphi_u^m = \left(\frac{i}{\pi}\right)^m m!\ \det(\xi_{w_\mu} \mid \xi_{w_\nu})\, dw_1 \wedge d\bar{w}_1 \wedge \cdots \wedge dw_m \wedge d\bar{w}_m$$

$$\varphi_\alpha \upsilon_\alpha = \varphi_u^m = 2^m m!\ |(\xi_{w_1} \circ \alpha) \wedge \cdots \wedge (\xi_{w_m} \circ \alpha)|^2 \upsilon_\alpha$$

$$\varphi_\alpha = 2^m\, m!\, |(\mathfrak{r}_{w_1} \circ \alpha) \wedge \dots \wedge (\mathfrak{r}_{w_m} \circ \alpha)|^2$$

$$\mathrm{Ric}\ \kappa = \frac{1}{4\pi} dd^c \log \varphi_\alpha = \frac{1}{4\pi} dd^c \log |\mathfrak{r}_{w_1} \circ \alpha \wedge \dots \wedge \mathfrak{r}_{w_m} \circ \alpha|^2$$

$$= \Delta^*(\ddot{\omega}_{m-1}) \geqq 0$$

Hence

$$\mathrm{Ric}_u(r) = \mathrm{Ric}_u^+(r) = \int_0^r \int_{G_t} \mathrm{Ric}\ \kappa \wedge \varphi^{m-1}\, dt$$

$$= \int_0^r \int_{M(e^t)} \Delta^*(\ddot{\omega}_{m-1}) \wedge \varphi^{m-1}\, dt$$

$$= \int_1^{e^r} \int_{M(r)} \Delta^*(\ddot{\omega}_{m-1}) \wedge \varphi^{m-1} \frac{dt}{t}$$

$$= \hat{T}_\Delta(e^r,1) = \hat{T}_\Delta(e^r,0) = \hat{T}_\Delta(e^r)$$

q.e.d.

Hence the Ricci function on M for u is the characteristic function of the Gauss map of M.

Now, B will be introduced by the following general assumptions.

(C5) Let B be a holomorphic form of bidegree (m-1,0) on M such that f is general for B. If m=1, take $B \equiv 1$.

(C5') Let B^0 be a holomorphic form of bidegree (m-1,0) on V. Assume that the coefficients of B^0 are polynomials of atmost degree n-1. Define $B = \mathfrak{w}^*(B^0)$. Assume that f is general for B.

If $m = 1$, take $B^0 \equiv 1$.

Obviously, (C5') implies (C5). If (C1) - (C3) are assumed, Theorem 7.11 shows that (C5') can be satisfied. The assumptions (C1) - (C5) imply (B1) - (B9) with the indicated choices. Hence the associated maps

$$f_p: M \longrightarrow \mathbb{P}(\textstyle\bigwedge_{p+1} V)$$

are defined. Write

$$n_M^p(r) = \hat{A}_{f_p}(r) = \int_{M(r)} f_p^*(\ddot{\omega}_p) \wedge \omega^{m-1} = A_{f_p}(\log^+ r)$$

$$N_M^p(r) = \hat{T}_{f_p}(r) = \int_0^r n_{M,p}(t) \frac{dt}{t} = T_{f_p}(\log^+ r)$$

$$Y(r) = Y_u(\log^+ r)$$

$$Z(r) = 2r^2\, Y(r) = Z_u(\log^+ r)$$

<u>Lemma 14.5</u>. Assume (C1) - (C4) and (C5'). Assume $m > 1$. Then a constant $c \geqq 1$ exists such that

$$Y(r) \leqq c\, r^{\frac{2n-2}{m-1}} \qquad \text{for all } r \geqq 1.$$

Proof. Define $q = m-2$. Let $\mathfrak{T} = \mathfrak{T}(q,n)$ be the set of all injective increasing maps $\nu: \mathbb{Z}[0,q] \to \mathbb{Z}[0,n]$. Take an orthonormal base $\mathfrak{w}_0, \ldots, \mathfrak{w}_n$ of V. Let $z_0, \ldots, z_n$ be the coordinate function of V for this base. For $\nu \in \mathfrak{T}$, define

$$\zeta_\nu = dz_{\nu(0)} \wedge \cdots \wedge dz_{\nu(q)}$$

$$\xi_\nu = \left(\frac{i}{2\pi}\right)^{m-1} dz_{\nu(0)} \wedge d\overline{z}_{\nu(0)} \wedge \cdots \wedge dz_{\nu(q)} \wedge d\overline{z}_{\nu(q)}$$

Then

$$\xi_\nu = i_0 \, \zeta_\nu \wedge \overline{\zeta}_\nu \geqq 0$$

$$\upsilon^{m-1} = (m-1)! \sum_{\nu \in \mathfrak{J}} \xi_\nu$$

$$B^0 = \sum_{\nu \in \mathfrak{J}} B_\nu \, \zeta_\nu$$

Here B_ν is a polynomial of atmost degree n-1 on V. Therefore a constant $c_0 > 0$ exists such that

$$|B_\nu|^2 \leqq c_0 \, \tau_0^{n-1} \qquad \text{for all } \nu \in \mathfrak{J}$$

on V - V[1]. Take $\nu \in \mathfrak{J}$ and $\mu \in \mathfrak{J}$. Then

$$0 \leqq i_0 (B_\mu \zeta_\mu - B_\nu \zeta_\nu) \wedge (\overline{B}_\mu \overline{\zeta}_\mu - \overline{B}_\nu \overline{\zeta}_\nu)$$

$$= i_0 \, [\, |B_\mu|^2 \zeta_\mu \wedge \overline{\zeta}_\mu - B_\mu \overline{B}_\nu \zeta_\mu \wedge \overline{\zeta}_\nu - B_\nu \overline{B}_\mu \zeta_\nu \wedge \overline{\zeta}_\mu + |B_\nu|^2 \zeta_\nu \wedge \overline{\zeta}_\nu]$$

Hence

$$i_0 \, (B_\mu \overline{B}_\nu \zeta_\mu \wedge \overline{\zeta}_\nu + B_\nu \overline{B}_\mu \zeta_\nu \wedge \overline{\zeta}_\mu) \leqq |B_\mu|^2 \, \xi_\mu + |B_\nu|^2 \, \xi_\nu$$

Therefore

$$i_0\, B^0 \wedge \overline{B}^0 = i_0 \sum_{\mu\in\mathfrak{z}} \sum_{\nu\in\mathfrak{z}} B_\mu \overline{B}_\nu \zeta_\mu \wedge \overline{\zeta}_\nu$$

$$= \frac{i_0}{2} \sum_{\mu\in\mathfrak{z}} \sum_{\nu\in\mathfrak{z}} (B_\mu \overline{B}_\nu \zeta_\mu \wedge \overline{\zeta}_\nu + B_\nu \overline{B}_\mu \zeta_\nu \wedge \overline{\zeta}_\mu)$$

$$\leqq \frac{1}{2} \sum_{\mu\in\mathfrak{z}} \sum_{\nu\in\mathfrak{z}} (|B_\mu|^2\, \xi_\mu + |B_\nu|^2\, \xi_\nu)$$

$$= \binom{n+1}{m-1} \sum_{\mu\in\mathfrak{z}} |B_\mu|^2\, \xi_\mu$$

$$\leqq \frac{1}{(m-1)!} \binom{n+1}{m-1} c_0\, \tau_0^{n-1}\, \upsilon^{m-1}$$

$$\leqq (2c)^{m-1}\, \tau_0^{n-1}\, \upsilon^{m-1}$$

on $V - V[1]$, where $c > 1$ is a constant. Consider the pull-back under $\mathfrak{w} : M \to V$. Then

$$i_0\, B \wedge \overline{B} \leqq (c\, r^{\frac{2n-2}{m-1}})^{m-1}\, \varphi_u^{m-1}$$

on $M[r]$. Therefore

$$Y(r) = Y_u(\log^+ r) \leqq c\, r^{\frac{2n-2}{m-1}} \qquad \text{for } r > r_1$$

$$Y(r) = Y_u(\log^+ r) = 1 \leqq c\, r^{\frac{2n-2}{m-1}} \qquad \text{for } r_1 \geqq r \geqq 1$$

q.e.d.

Observe

$$Z(r) = 2r^2\, Y(r) \leqq 2c\, r^{\frac{2n-2}{m-1} + 2}$$

If $m = 1$, then $Y(r) \equiv 1$ and $Z(r) = 2r^2$.

If $0 \leqq p < n$, then

$$H_p = i_0\, B \wedge \overline{B} \wedge f_p(\omega_p) = h_p\, \varphi^m = h_{up}\, \varphi_u^m$$

Hence the <u>inflection functions</u> are given by

$$S_M^p(r) = \frac{1}{2} \int_{M(r)} \log \frac{1}{h_p}\, \omega^m = S_f^p(\log^+ r)$$

$$\overset{+}{S}{}_M^p(r) = \frac{1}{2} \int_{M(r)} \log^+ \frac{1}{h_p}\, \omega^m = \overset{+}{S}{}_f^p(\log^+ r)$$

$$s_M^p(r) = \frac{1}{2} \int_{M(r)} \log \frac{1}{h_{up}}\, \omega^m = S_f^p(\log^+ r; u)$$

$$\overset{+}{s}{}_M^p(r) = \frac{1}{2} \int_{M(r)} \log^+ \frac{1}{h_{up}}\, \omega^m = S_f^p(\log^+ r; u)$$

$$\overset{+}{s}_M(r) = \sum_{p=0}^{n-1} \overset{+}{s}{}_M^p(r) = \overset{+}{S}_f(\log^+ r; u)$$

The deficit estimates imply Bezout estimates by the inequality

$$N_M(r;a) \leqq N_M(r) + D_M(r;a)$$

The transformation $u\colon \mathbb{R}[0,+\infty) \to \mathbb{R}[1,+\infty)$ sends a set of finite measure into a set E of <u>finite logarithmic measure</u> on $\mathbb{R}[1,+\infty)$, i.e.

$$\int_E \frac{dt}{t}$$

If $h = g$ or $h \leqq g$ holds except for a set of finite logarithmic measure write $h \mathrel{\overset{\cdot\cdot}{=}} g$ respectively $h \mathrel{\overset{\cdot\cdot}{\leqq}} g$.

<u>Theorem 14.6</u>. <u>First Asymptotic Bezout Estimate</u>.
Assume (C1) - (C5). Then the following estimate holds for $a \in \mathbb{P}(V^*)$ and all $r > r_1$ except for a set of finite logarithmic measure.

$$N_M(r;a) \underset{\cdot\cdot}{\leqq} N_M(r) + 2^n n_M(r) \sum_{p=0}^{n-1} \log \left(1 + \frac{N_M^p(r)}{n_M(r)}\right)$$

$$+ n\, 2^{n-1} n_M(r) [\log^+ n_M(r) + (m-1)\log Y(r) + 3m\log r]$$

$$+ 2^n \sum_{p=0}^{n-1} \overset{+}{S}{}_M^p(r)$$

If (C5') holds, there exists a constant $c_0 > 0$ depending on n, m and B^0 only such that

$$N_M(r,a) \underset{\cdot\cdot}{\leqq} N_M(r) + 2^n n_M(r) \sum_{p=0}^{n-1} \log \left(1 + \frac{N_M^p(r)}{n_M(r)}\right)$$

$$+ n\, 2^{n-1} n_M(r) (\log^+ n_M(r) + c_0 \log r)$$

$$+ 2^n \sum_{p=0}^{n-1} \overset{+}{S}{}_M^p(r)$$

Proof. Apply Theorem 11.6 with the substitution $r \to \log^+ r$. Then

$$c + (m-1) \log^+ Z_u(\log^+ r) = c + (m-1) \log^+(2r^2 Y(r))$$

$$\underset{\cdot\cdot}{\leqq} 3m \log r + (m-1) \log^+ Y(r)$$

If (C5') holds, then $3m \log r + (m-1) \log^+ Y(r) \leqq c_0 \log r$

q.e.d.

Theorem 14.7. Second Asymptotic Bezout Estimate. Assume (C1) - (C5). Then a constant $C > 0$ exists such that for all $a \in \mathbb{P}(V^*)$ and all $r > 1$ except a set of finite logarithmic measure the following estimate holds

$$N_M(r;a) \underset{\cdot\cdot}{\leqq} N_M(r) + 2^n \overset{+}{s}_M(r) + C\, n_M(r) \log \overset{+}{s}_M(r)$$

$$+ C\, n_M(r) \,[\log^+ N_M(r) + \log^+ \hat{T}_\Delta(r) + \log Y(r)$$

$$+ \log r]$$

If also (C5') holds, a constant $C_1 > 0$ exists such that

$$N_M(r,a) \underset{\cdot\cdot}{\leqq} N_M(r) + 2^n \overset{+}{s}_M(r) + C\, n_M(r) \log \overset{+}{s}_M(r)$$

$$+ C\, n_M(r) \,[\log^+ N_M(r) + \log^+ \hat{T}_\Delta(r) + C_1 \log r]$$

Proof. Apply Theorem 12.10 with the substitution $r \to \log^+ r$. Observe $Q_u(r) = 0$ and $\mathrm{Ric}_u^+(\log r) = \hat{T}_\Delta(r)$. Also

$$\log^+ Z_u(\log r) = \log 2r^2 Y(r) = \log Y(r) + 2 \log r + 2$$

$$\leqq \log Y(r) + 3 \log r$$

Also $n_M(r) \underset{\cdot\cdot}{\leqq} N_M(r)^2$ or $\log^+ n_M(r) \underset{\cdot\cdot}{\leqq} 2 \log^+ N_M(r)$. These estimates and Theorem 12.10 imply the first estimate.

If (C5') holds then $Y(r) \leqq C_0\, r^{\frac{2n-2}{m-1}}$. Hence

$$\log r + \log Y(r) \leqq c_1 \log r$$

which implies the second estimate.

q.e.d.

The manifold M is said to be steady for B if and only if f is steady for B which is the case if and only if $h_{up} \geqq C > 0$ on M for some positive constant which means $H_p \geqq C\, \varphi_u^m > 0$ on M.

Theorem 14.8. Third Asymptotic Bezout Estimate.

Assume (C1) - (C5). Assume that M is steady for B. Then there exists a constant $C > 0$ such that for all $r > 1$ except a set of finite logarithmic measure and for all $a \in \mathbb{P}(V^*)$ the inequality

$$N_M(r;a) \underset{\bullet\bullet}{\leqq} N_M(r)+C\, n_M(r)\, [\log^+ N_M(r)+\log^+ \hat{T}_\Delta(r)+\log r + \log Y(r)]$$

holds. If also (C5') is assumed, then a constant $C_1 > 0$ exists such that for all $r > 1$ except a set of finite logarithmic measure and for all $a \in \mathbb{P}(V^*)$ the following inequality holds

$$N_M(r;a) \underset{\bullet\bullet}{\leqq} N_M(r) + C_1\, n_M(r)\, [\log^+ N_M(r) + \log^+ \hat{T}_\Delta(r) + \log r]$$

Proof. Theorem 14.2 and Theorem 13.4 imply

$$N_M(r;a) \underset{\bullet\bullet}{\leqq} N_M(r) + C\, n_M(r)\, [\log^+ N_M(r) + \log^+ 2r^2\, Y(r)$$

$$+ \log^+ n_M(r) + \log^+ \hat{T}_\Delta(r)]$$

since $Q_u \equiv 0$. Now $n_M(r) \leqq N_M(r)^2$ and const $\leqq \log r$ imply the first assertion with another constant C. If also (C5') is assumed, then

$$\log^+ Y(r) \leqq C_0 + \frac{2n-2}{m-1} \log^+ r \leqq \tilde{C} \log r$$

which implies the second assertion; q.e.d.

Now, uniform estimates shall be considered.

Theorem 14.9. First Uniform Bezout Estimate. Assume (C1) - (C5). Then there exists a constant $C_0 > 0$ depending on n and m only such that for all ε with $0 < \varepsilon < 1$ for all $a \in \mathbb{P}(V^*)$ and all $r > r_1$ the following inequality holds

$$N_M(r;a) \leqq N_M(r) + 2^n n_M(r) \sum_{p=0}^{n-1} \log^+ \frac{N_M^p(r+\varepsilon) + n_M(r+\varepsilon)}{n_M(r) \log(1+\frac{\varepsilon}{r})}$$

$$+ n_M(r) [C_0 + n(m-1) 2^n \log(r\, Y(r+\varepsilon))]$$

$$+ 2^n \sum_{p=0}^{n-1} \overset{+}{S}{}_M^p(r)$$

If also (C5') is satisfied, then there exists a constant $C_1 > 0$ depending on n, m and B^0 only such that for all ε in $0 < \varepsilon < 1$ for all $a \in \mathbb{P}(V^*)$ and all $r > r_1$ the following inequality holds

$$N_M(r;a) \leqq N_M(r) + 2^n n_M(r) \sum_{p=0}^{n-1} \log^+ \frac{N_M^p(r+\varepsilon) + n_M(r+\varepsilon)}{n_M(r) \log(1+\frac{\varepsilon}{r})}$$

$$+ C_1 \, n_M(r) \log r + 2 \sum_{p=0}^{n-1} \overset{+}{S}{}^p_M(r)$$

Proof. Replace r by log r and θ by $\theta = \dfrac{\log (r+\varepsilon)}{\log r} > 1$ in Theorem 11.5 with $r > r_1 > 1$. Then

$$\theta \log r = \log (r + \varepsilon)$$

$$(\theta-1) \log r = \log \left(1 + \frac{\varepsilon}{r}\right)$$

$$u'(\theta \log r) = u'(\log(r + \varepsilon)) = 2(r + \varepsilon)^2 \leqq 8r^2$$

$$Y_u(\theta \log r) = Y_u(\log(r + \varepsilon)) = Y(r + \varepsilon) \leqq Y(r + \varepsilon)^2$$

Hence

$$\log^+ Z_u(\theta \log r) \leqq \log 8 \, r^2 \, Y(r+\varepsilon)^2 = 2 \log(rY(r+\varepsilon)) + \log 8$$

Define $c_0 = c + n(m-1) \log 8$. Substitution proves the first asserted inequality. If also (C5') holds, then a constant $C_1 > 0$ exists depending on n, m and B^0 only such that

$$c_0 + n(m-1) \, 2^n \log (r \, Y(r+\varepsilon))$$

$$\leqq c_0 + n(m-1) \, 2^n \log (r \, Y(r+1)) \leqq C_1 \log r$$

by Lemma 14.5 where C_1 depends on n, m and B^0 only. This estimate implies the second assertion.

Theorem 14.10. Second Uniform Bezout Estimate. Assume (C1) - (C5). Then there exists a constant $C_0 > 0$ depending only on n and m such that for all ε with $0 < \varepsilon < 1$, for all $a \in \mathbb{P}(V^*)$ and all $r > e^2$, the following inequality holds

$$N_M(r;a) \leqq N_M(r) + 2^n + 2^n \overset{+}{s}_M(r) +$$

$$2^{2n+1} n_M(r) [2 \log N_M(r+\varepsilon) + \log^+ \hat{T}_\Delta(r+\varepsilon) +$$

$$\log^+ \overset{+}{s}_M(r+\varepsilon) + (m-1) \log^+ Y(r+\varepsilon) + (2m+1)\log r + C_0 + 2\log \frac{1}{\varepsilon}$$

If also (C5') is assumed, then there exists a constant $C_1 > 0$ depending on n, m and B^0 only such that for all ε with $0 < \varepsilon < 1$, for all $a \in \mathbb{P}(V^*)$ and all $r > e^2$ the following inequality holds

$$N_M(r,a) \leqq N_M(r) + 2^n + 2^n \overset{+}{s}_M(r)$$

$$+ 2^{2n+1} n_M(r) [2 \log^+ N_M(r+\varepsilon) + \log^+ \hat{T}_\Delta(r+\varepsilon) +$$

$$+ \log^+ \overset{+}{s}_M(r+\varepsilon) + C_1 \log r + 2 \log \frac{1}{\varepsilon}]$$

Proof. As in the proof of Theorem 14.9 take

$$\theta = \frac{\log (r+\varepsilon)}{\log r}$$

Replace r by log r with $r > e^2$ in Theorem 12.13. Then

$\log r > 2$ and

$$\frac{\theta}{\theta-1} = \frac{\log (r+\varepsilon)}{\log (1+\frac{\varepsilon}{r})} = 1 + \frac{\log r}{\log (1 + \frac{\varepsilon}{r})}$$

If $0 < x < e-1$, then

$$\log (1 + x) \geqq \frac{x}{e-1} \geqq \frac{1}{2} x$$

Now $0 < \frac{\varepsilon}{r} < \frac{1}{r} < 1 < e-1$. Hence $\log (1+\frac{\varepsilon}{r}) \geqq \frac{\varepsilon}{2r}$. Therefore

$$\frac{\theta}{\theta-1} \leqq 1 + \frac{2r}{\varepsilon} \log r \leqq \frac{4r}{\varepsilon} \log r \leqq \frac{4r^2}{\varepsilon}$$

$$\log \frac{\theta}{\theta-1} \leqq \log \frac{1}{\varepsilon} + \log 4 + 2 \log r$$

Also

$$\log^+ Z_u(\theta \log r) = \log [8r^2\, Y(r+\varepsilon)] = \log 8 + 2 \log r + \log Y(r+\varepsilon).$$

Observe

$$n_M(r+\varepsilon) \leqq (\log(1 + \frac{\varepsilon}{r}))^{-1} \int_r^{r+\varepsilon} n_M(t) \frac{dt}{t} \leqq \frac{2r}{\varepsilon} N_M(r+\varepsilon)$$

$$\log^+ n_M(r+\varepsilon) \leqq \log^+ N_M(r+\varepsilon) + \log 2 + \log r + \log \frac{1}{\varepsilon}$$

With these estimates and substitutions, Theorem 14.2 and Theorem 12.13 imply the first assertion. If also (C5') holds, then Lemma 14.5 implies

$$C_0 + (2m+1) \log r + (m-1) \log Y(r+\varepsilon)$$

$$\leqq C_0 + (2m+1) \log r + (m-1) \log Y(r+1) \leqq C_1 \log r$$

for some constant $C_1 > 0$ depending only on n, m and B^0, which implies the second assertion; q.e.d.

The last estimate in Theorem 14.10 can be simplified. Observe

$$N_M(r) = N_M(r,1) \leqq n_M(r) \log r$$

Let the constants C_ν depend on M, V and ε. If $r > r_1 + e^2$. Then $n_M(r+1) \geqq n_M(r_1+1) > 0$. Therefore a constant $C_2 > 0$ exists such that

$$2n_M(r) \log \frac{1}{\varepsilon} + 2^{-n-1} \leqq C_2\, n_M(r) \log r$$

Therefore

$$N_M(r,a) \leqq 2^{2n+1}\, n_M(r)\, [2 \log^+ N_M(r+\varepsilon) + \log^+ \hat{T}_\Delta(r+\varepsilon)$$

$$+ \log^+ \overset{+}{s}_M(r+\varepsilon) + (2^{-2n-1}+C_1+C_2) \log r] + 2^n\, \overset{+}{s}_M(r)$$

The following result is obtained.

<u>Corollary 14.11</u>. Assume (C1) - (C4) and (C5'). Take $0 < \varepsilon < 1$. Then there exists a constant $C > 0$ such that for all $a \in \mathbb{P}(V^*)$ and all $r > r_1 + e^2$, the following

estimate holds

$$N_M(r;a) \leqq C\, n_M(r)\, [\log^+ N_M(r+\varepsilon) + \log^+ \hat{T}_\Delta(r+\varepsilon) + \log r]$$

$$+ 2^{2n+1}\, n_M(r) \log^+ \overset{+}{s}_M(r+\varepsilon) + 2^n\, \overset{+}{s}_M(r)$$

Now, the case of a steady M shall be considered.

Theorem 14.12. Third Uniform Bezout Estimate. Assume (C1) - (C5). Assume that M is steady for B. Take ε with $0 < \varepsilon < 1$. Then constants $C_0 > 0$ and $C_1 > 0$ exist such that for all $a \in \mathbb{P}(V^*)$ and all $r > r_1 + e^2$, the following estimates hold

$$N_M(r;a) \leqq N_M(r) + C_0\, n_M(r) \log r +$$

$$+ 2^{2n+1} n_M(r)\, [3 \log^+ N_M(r+\varepsilon) + \log^+ \hat{T}_\Delta(r+\varepsilon) + (m-1) \log Y(r+\varepsilon)]$$

$$N_M(r;a) \leqq C_1\, n_M(r)\, [\log^+ N_M(r+\varepsilon) + \log^+ Y(r+\varepsilon) + \log^+ \hat{T}_\Delta(r+\varepsilon)]$$

If also (C5') is assumed, then a constant $C_2 > 0$ exists such that for all $a \in \mathbb{P}(V^*)$ and all $r > r_1 + e^2$, the following estimate holds

$$N_M(r;a) \leqq C_2\, n_M(r)\, [\log^+ N_M(r+\varepsilon) + \log^+ \hat{T}_\Delta(r+\varepsilon) + \log r]$$

Proof. A constant $C_3 > 0$ exists such that

$$\overset{+}{S}_f(\log r;u) \leqq C_3\ \Phi(\log r) \qquad \text{if } r > 1$$

Hence $\overset{+}{s}_M(r) \leqq C_3\ n_M(r)$ and

$$\log^+ \overset{+}{s}_M(r) \leqq \log^+ C_3 + \log^+ n_M(r+\varepsilon)$$

$$\leqq \log^+ C_3 + \log^+ (\tfrac{2r}{\varepsilon} N_M(r+\varepsilon))$$

$$\leqq C_4 \log r + \log^+ N_M(r+\varepsilon)$$

if $r > r_1 + e^2$ for some constant $C_4 > 0$. Also

$$2^n + 2^n\ \overset{+}{s}_M(r) \leqq 2^n + 2^n\ C_3\ n_M(r) \leqq C_5\ n_M(r) \log r$$

$$2^{2n+1}\ (2m+1) \log r + c_0 + 2 \log \frac{1}{\varepsilon} \leqq C_6 \log r$$

where C_5 and C_6 are constants. Theorem 14.10 implies

$$N_M(r;a) \leqq N_M(r) + 2^{2n+1}\ [3 \log^+ N_M(r+\varepsilon) + \log^+ \hat{T}_\Delta(r+\varepsilon) + (m-1) \log Y(r+\varepsilon)$$

$$+ [C_5 + 2^{2n+1}\ C_4 + C_6]\ n_M(r) \log r$$

which implies the first assertion with $C_0 = C_5 + 2^{2n+1} C_4 + C_6$.

Observe $N_M(r) \leqq n_M(r) \log r$ for $r > 1$, which implies the second assertion with some constant $C_1 > 0$. If (C5') is assumed then

$$\log^+ Y(r+\varepsilon) \leqq \frac{2n-2}{m-1} \log (r+1) + C_7 \leqq C_8 \log r$$

which implies the third assertion; q.e.d.

REFERENCES

[1] Ahlfors, L.: The theory of meromorphic curves. Acta. Soc. Sci. Fenn Nova Ser. A3 (4) (1941), p.31.

[2] Andreotti, A. and W. Stoll: Analytic and algebraic dependence of meromorphic functions. Lecture Notes in Mathematics 234. Springer-Verlag. Berlin-Heidelber-New York, 1971, p. 390.

[3] Bloom, T. and M. Herrera: De Rham cohomology of an analytic space. Invent. Math. 7 (1969), p. 275-296.

[4] Bott, R. and S. S. Chern: Hermitian vector bundles and the equidistribution of the zeroes of their holomorphic sections. Acta Math. 114 (1965) p. 71-112.

[5] Carlson, J.: Some degeneracy theorems for entire functions with values in an algebraic variety. Transactions Amer. Math. Soc. 168 (1972), p. 273-301.

[6] Carlson, J. and Ph. Griffiths: A defect relation for equidimensional holomorphic mappings between algebraic varieties. Ann. of Math. (2) 95 (1972), p. 557-584.

[7] Chern, S. S.: The integrated form of the first main theorem for complex analytic mappings in several variables. Ann. of Math. (2) 71 (1960), p. 536-551.

[8] Chern S. S.: Holomorphic curves in the plane. Diff. Geom., in honor of K. Yano, Kinokuniya, Tokyo, 1972, p. 73-94.

[9] Cornalba, M. and B. Shiffman: A counter example to the "transcendental Bezout Problem". Ann. of Math. (2) 96 (1972), p. 402-406.

[10] Cowen, M.: Value distribution theory. M.I.T. Thesis, 1971, p. 90 (to appear in Transactions Amer. Math. Soc.)

[11] Dektjarev, I.: The general first fundamental theorem of value distribution theory. Dokl. Akad. Nank. SSR 193 (1970) (Soviet Math. Dokl. 11 (1970) p. 961-963).

[12] Federer, H.: Geometric measure theory. Die Grundl. d. Math. Wiss. 153 Springer-Verlag, Berlin-Heidelberg-New York 1969 p. 676.

[13] Green, M.: Holomorphic maps into complex projective space omitting hyperplanes. Transactions Amer. Math. Soc. 169 (1972) p. 89-103.

[14] Griffiths, Ph.: Holomorphic mapping into canonical algebraic varieties. Ann. of Math. (2) 93 (1971), p. 439-458.

[15] Griffiths, Ph.: Function theory of finite order on algebraic varieties I. Jour. of Diff. Geom. 6 (1972) p. 285-306.

[16] Griffiths, Ph.: On the Bezout problem for entire analytic sets. p. 37 (to appear).

[17] Griffiths, Ph.: Two results in the global theory of holomorphic mappings. p. 39 (to appear in Contributions in Analysis in honor of L. Bers).

[18] Griffiths Ph. and J. King: Nevanlinna theory and holomorphic mappings between algebraic varieties. Acta Mathematica 130 (1973) p. 145-220.

[19] Herrera, M.: Integration on a semi-analytic set. Bull. Soc. Math. France 94 (1966), p. 141-180.

[20] Hirschfelder, J.: The first main theorem of value distribution in several variables. Invent. Math. 8 (1969) p. 1-33.

[21] Kiernan, P. and Kobayashi, S.: Holomorphic mappings into projective space with lacunary hyperplanes. p. 31 (to appear).

[22] King, J.: The currents defined by analytic varieties. Acta Math. 127 (1971), p. 185-220.

[23] Kneser, H.: Zur Theorie der gebrochenen Funktionen mehrerer Veränderlichen. Jber. dtsch. Math. Ver. 48 (1938), p. 1-38.

[24] Lelong, P.: Intégration sur une ensemble analytique complexe. Bull. Soc. Math. France 85 (1957), p. 328-370.

[25] Lelong, P.: Fonctions pluri-sous-harmoniques et formes differentielles positives. Gordon and Breach. (1968) p. 301.

[26] Levine, H.: A theorem on holomorphic mappings into complex projective space. Ann. of Math. (2) 71 (1960), p. 529-535.

[27] Nevannlina, R.: Eindeutige analytische Funktionen. Die Grundl. d. Math. Wiss. XLVC. Springer-Verlag Berlin-Göttingen-Heidelberg 2. ed. 1953, p. 379.

[28] Remmert, R.: Holomorphe und meromorphe Abbildungen komplexer Räume. Math. Ann. 133 (1957), p. 338-370.

[29] Sard, A.: The measure of the critical values of differentiable maps. Bull. Amer. Soc. 48 (1972), p.883-890.

[30] Schwartz, M.-H.: Formules apparentées a la formule de Gauss-Bonnet pour certaines applications d'une variété a n dimensions dans une autre. Acta Math. 91 (1954) p. 189-244.

[31] Schwartz, M.-H.: Formules apparentées a celles de Nevanlinna-Ahlfors pour certaines applications d'une variété a n dimensions dans une autre. Bull Soc. Math. France 82 (1954), p. 317-360.

[32] Shiffman, B.: Extension of positive line bundles and meromorphic maps, Invent. Math. 15 (1972), p. 332-347.

[33] Stein, K.: Maximale holomorphe und meromorphe Abbildungen I. Amer. J. of Math. 85 (1963) p. 298-313. II ibid. 86 (1964) p. 823-868.

[34] Stoll, W.: Mehrfache Integrale auf komplexen Mannigfaltigkeiten. Math. Zeitschr. 57, (1952), p. 116-154.

[35] Stoll, W.: Ganze Funktionen endlicher Ordnung mit gegebenen Nullstellenflächen. Math. Zeitschr. 57 (1953), p. 211-237.

[36] Stoll, W.: Die beiden Hauptsätze der Wertverteilungstheorie bei Funktionen mehrerer komplexer Veränderlichen I. Acta Math. 90 (1953) p. 1-115. II ibid. 92 (1954), p. 55-169.

[37] Stoll, W.: The growth of the area of a transcendental analytic set. I. Math. Ann. 156 (1964), p. 47-48. II. ibid. 156 (1964), p. 144-170.

[38] Stoll, W.: Normal families of non-negative divisors. Math. Zeitschr. 84 (1964), p. 154-218.

[39] Stoll, W.: A general first main theorem of value distribution. Acta Math. 118 (1967), p. 111-191.

[40] Stoll, W.: About entire and meromorphic functions of exponential type. Proceed. of Symposia in Pure Math. 11 (1968), p. 392-430.

[41] Stoll, W.: About the value distribution of holomorphic maps into projective space. Acta Math. 123 (1969), p. 83-114.

[42] Stoll, W.: Value distribution of holomorphic maps into compact complex manifolds. Lecure Notes in Mathematics 135. Springer-Verlag. Berlin-Heidelberg-New York 1970, p. 267.

[43] Stoll, W.: Value distribution of holomorphic maps. Several Complex Variables I. Maryland 1970. Lecture Notes in Mathematics 155 (1970) p. 165-190. Springer-Verlag. Berlin-Heidelberg-New York.

[44] Stoll, W.: Fiber integration and some of its applications. Symposium on Several Complex Variables. Park City, Utah, 1970. Lecture Notes in Mathematics 184 (1971) p. 109-120. Springer-Verlag, Berlin-Heidelberg-New York.

[45] Stoll, W.: A Bezout estimate for complete intersections. Ann. of Math. (2) 96 (1972) p. 361-401.

[46] Thie, P.: The Lelong number of a point of a complex analytic set. Math. Ann. 172 (1967), p. 269-312.

[47] Tung, Ch.: The first main theorem on complex spaces. (1973 Notre Dame Thesis).

[48] Weyl, H. and J. Weyl: Meromorphic functions and analytic curves. Annals of Math. Studies 12. Princeton Univ. Press, Princeton, N.J. 1943, p. 269.

[49] Wu, H.: Mappings of Riemann surfaces (Nevanlinna theory). Proceed. of Symposia in Pure Math. 11 (1968) p. 480-532.

[50] Wu, H.: Remarks on the first main theorem of equidistribution theory I, Jour. of Diff. Geom. 2 (1968) p. 197-202. II ibid. 2 (1968), p. 369-384. III ibid. 3 (1969) p. 83-94. IV ibid. (1969), p. 433-446.

[51] Wu, H.: The equidistribution theory of holomorphic curves. Annals of Math. Studies 64, Princeton Univ. Press, Princeton, N.J. 1970, p. 219.

INDEX OF GENERAL ASSUMPTIONS

(A1) Let M be a connected complex manifold of dimension m.

(A2) Let $\chi \geqq 0$ be a non-negative, differential form of class C^∞ and of bidegree (m-1,m-1) on M such that $d\chi = 0$.

(A3) Let V be a hermitian vector space of dimension n+1 with $n > 0$.

(A4) Let $f: M \to \mathbb{P}(V)$ be a meromorphic map.

(A5) Let B be a holomorphic differential form of bidegree (m-1,0) on M. Let $h'_\alpha = D_{B,\alpha}h$ be the associated, contravariant differential operator. (If m = 1, take $B \equiv 1$).

(A6) Assume that f is general for B:

(A7) Assume that a bump (G,g,ψ) on M is given. Assume that $i_0 B \wedge \overline{B} \leqq \chi$ where i_0 is defined by (8.1).

(B1) Let M be a non-compact, connected, complex manifold of dimension $m > 0$.

(B2) Let τ be a pseudo-convex exhaustion of M. Define

$$r_0 = \inf \{r \in \mathbb{R}[0,+\infty) \mid G_r \neq \emptyset\}$$

(B3) Let V be a hermitian vector space of dimension n+1 with $n > 0$.

(B4) Let $f: M \longrightarrow \mathbb{P}(V)$ be a meromorphic map.

(B5) Let B be a holomorphic differential form of bidegree (m-1,0) on M. Let $h' = h'_\alpha = D_{B,\alpha}h$ be the associated contravariant differential operator.

(B6) Assume that f is general for B.

(B7) A convex change of scale is given, which majorizes B. Define

$$Z_u = Y_u \cdot u' > 0$$

(B8) A strongly convex change of scale u of τ is given on M. (B8 implies B7).

(B9) Let κ be the hermitian metric along the fibers of the canonical bundle K on M defined by φ_u^m where u is given by (B8).

(B10) Assume that f is steady for B.

(B11) The meromorphic map grows sufficiently.

(B12) The meromorphic map grows slowly.

(C1) Let V be a hermitian vector space of dimension n + 1, with n > 0. Define $\tau_0: V \to \mathbb{R}$ by $\tau_0(\mathfrak{z}) = |\mathfrak{z}|^2$ if $\mathfrak{z} \in V$.

(C2) Let M be a connected, closed, smooth, complex submanifold of dimension m > 0 of V with M[1] = $\emptyset$. Let $\mathfrak{w}: M \to V$ be the inclusion map.

(C3) Assume that M is not contained in any proper linear subspace of V.

(C4) Take $u \in \mathfrak{U}$ with $u(r) = e^{2r}$ as a strong convex change of scale τ.

(C5) Let B be a holomorphic form of bidegree (m-1,0) on M such that f is general for B. If m=1, take B $\equiv$ 1.

(C5') Let B^0 be a holomorphic form of bidegree (m-1,0) on V. Assume that the coefficients of B^0 are polynomials of atmost degree n-1. Define $B = \mathfrak{w}^*(B^0)$. Assume that f is general for B. If m = 1, take $B^0 \equiv 1$.

INDEX

Wilhelm Stoll has been Professor of Mathematics at the University of Notre Dame since 1960. From 1954 to 1960 he was a Dozent at the University of Tübingen, where he received his Ph.D. in 1953. Dr. Stoll was also a postdoctoral student at the Federal Institute of Technology in Zurich.

The author's research interests concern complex analysis in several variables and most of his publications deal with this area of mathematics. He has been a visiting lecturer at the University of Pennsylvania (1954-55), a temporary member of the Institute for Advanced Study (1957-59), and a Visiting Professor at Stanford University (1968-69) and at Tulane University (1973).